PLANES and MODEL KITS

CURTISS P-40 WARHAWK

ASSEMBLY – TRANSFORMATION – PAINTING – WEATHERING

Emmanuel PERNES and Olivier SOULLEYS
Profiles by Nicolas GOHIN

CONTENTS

HISTOIRE & COLLECTIONS

CURTISS P-40

The XP-40 in its original form, immediately distinguishable with its radiator moved to the back. The aircraft is already close to first production P-40s.

After a first volume dedicated to the FOCKE WULF Fw 190, the collection *Planes and model kits* continues today with an often underestimated plane, the Curtiss P-40.

The reason for this is straightforward. Despite its limitations, the Warhawk was a plane that served throughout the whole war and on every front, bearing the most colourful insignia and nose art of the Second World War.

What could be more motivating for a model enthusiast than to make a plane with the most spectacular camouflage scheme!

This book does not pretend to have a historical approach, but offers an opportunity to discover a plane purely from a model maker's point of view. Each of the four kits is structured around a central technical theme, with the objective of presenting, in the most dynamic and explanatory way possible, some techniques used by model making experts.

These articles do not pretend to have the definitive answers, but rather to offer you a range of possible solutions to problems that can be encountered when putting together a kit. This collection will soon see the addition of new volumes and will look at other periods in aviation, from the First World War to the supersonic age.

We wish you a pleasant read.

The Curtiss P-40, a plane of duty

Encouraged by the performances of aircraft with in-line engines on the European front, Curtiss decided to develop a plane of this type by adapting an Allison V 1710-19 engine to the airframe of its piston engined P-36 fighter. Thus, the Curtiss H-81 or P-40 was created for the American air force. This plane would be handicapped throughout its career by the initial choices of the Curtiss engineers. Indeed, the airframe, despite its toughness, had reached its limits, whilst the Allison engine performed poorly at combat altitude, notably above 4,500 metres. These prohibitive faults meant it did not acquire the aura of the thoroughbred fighters such as the North American P-51 Mustang or the legendary status of the Spitfire and Messerschmitt Bf 109.

The P-40 is more associated with the period of Allied defeats at the beginning of the war. Crushed at Pearl Harbor without even taking off, completely dominated by the Bf 109F aircraft of JG27 in the North African desert, decimated by Zeroes during air raids on Darwin, the Warhawk did, nonetheless, have the merit of being present and putting up a fight against the triumphant onslaught of the Axis

A training flight over the United States with a P-40K, a P-40F and a P-40L in the background
(USAF)

forces, allowing the Allies to wait for better days.

The P-40 is the only aircraft that can pride itself on having seen action on all the fronts of the Second World War. It was used in the defence of American territory and the Panama Canal, in the Pacific at Pearl Harbor or Midway, in the extreme climate of the Aleutian Islands and the Russian front, and defended Australia and New Guinea against Japanese air raids. It was in North Africa that the P-40 got its first shark teeth nose art with which it is so often associated. It was also the P-40 that saw the reconstitution of the famous Escadrille La Fayette within the new French air force, with its famous Sioux Indian head, symbol of Franco-American friendship, allowing France to regain its lost honour.

However, it was in the Far East with General Claire Lee Chennault's Flying Tigers that the P-40 saw its only spectacular successes. This group of volunteers (or mercenaries) saw off the Japanese air force and allowed several pilots to become aces. The American Volunteer Group, the unit's official name, using combat tactics that well suited the aircraft's capacities, allowed the P-40 to enter, after the event, into the pantheon of warplanes. Despite its limitations, more than 14,000 Curtiss P-40 aircraft were made, all versions included. And although it rather under-performed as a fighter, it was, on the other hand, well disposed for ground attack; an area where its toughness came into its own.

Having gone through the whole war, it was therefore, one of the Second World War's key planes.

THE DIFFERENT VERSIONS OF THE P-40

P-40B & C

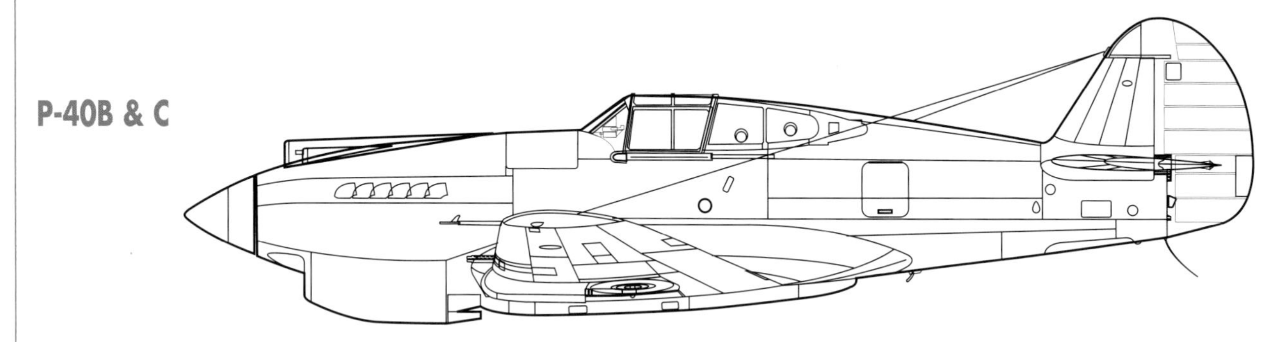

Identical on the outside, the P-40C received self-sealing fuel tanks. It was the Allison V 1710-33 that equipped these two versions.

P-40E

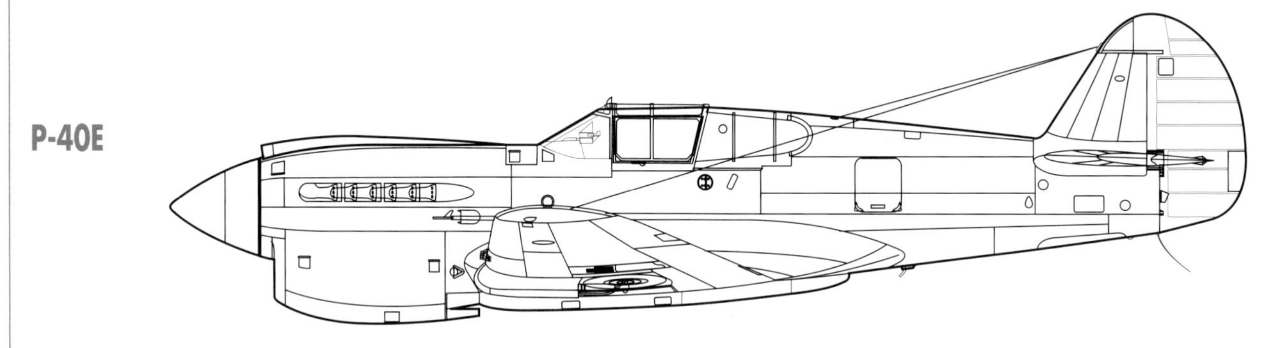

the P-40D was identical but only had four wing machine guns. The engine was now an Allison V 1710-33. The new engine and the removal of the nose guns led to a redesign of the aircraft's nose.

P-40F

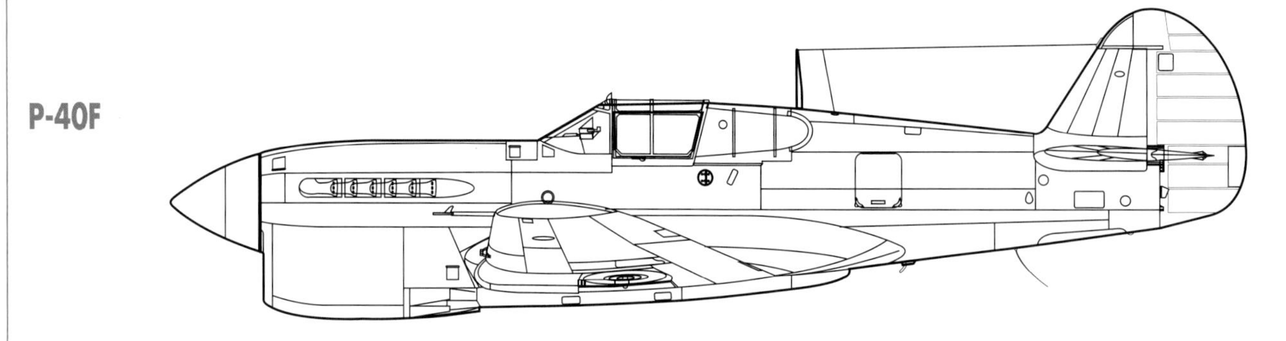

A Packard engine was adapted to try and resolve the problems linked to the Allison engine. Two types of airframe, short or long tails.

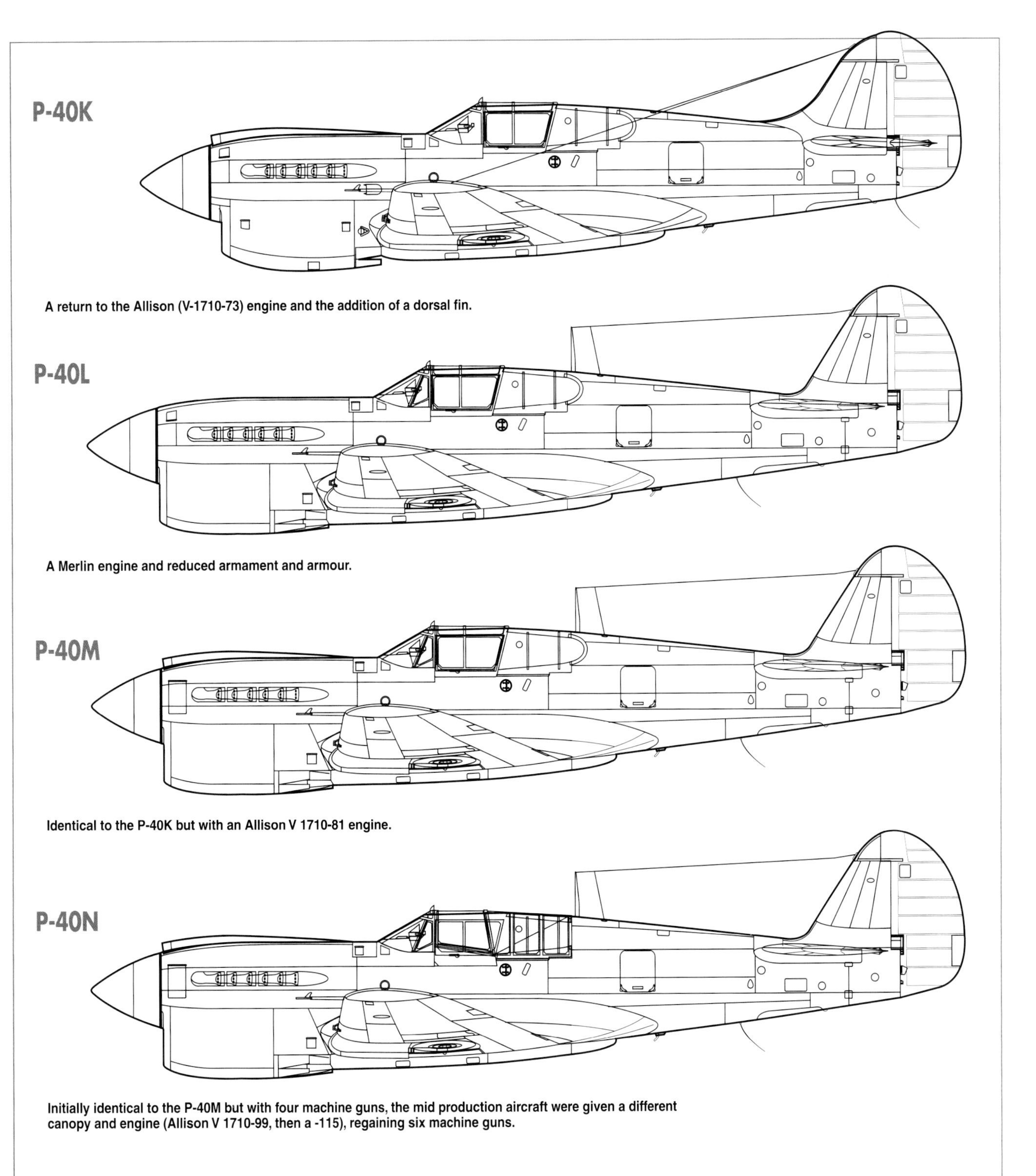

A return to the Allison (V-1710-73) engine and the addition of a dorsal fin.

A Merlin engine and reduced armament and armour.

Identical to the P-40K but with an Allison V 1710-81 engine.

Initially identical to the P-40M but with four machine guns, the mid production aircraft were given a different canopy and engine (Allison V 1710-99, then a -115), regaining six machine guns.

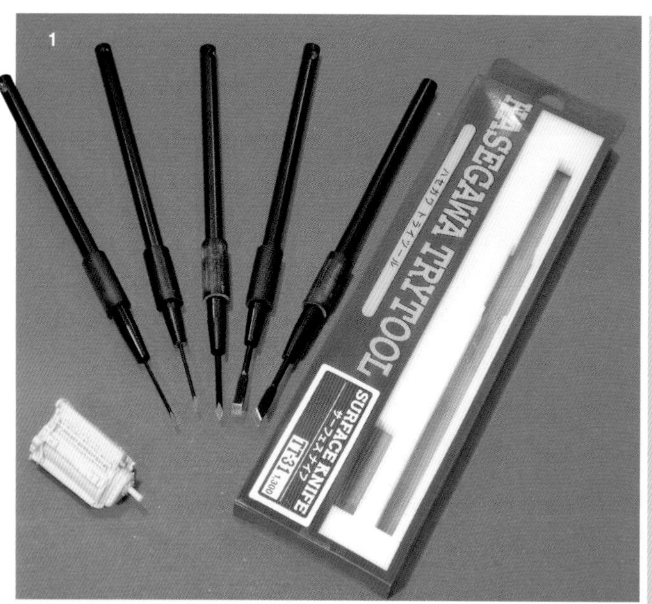

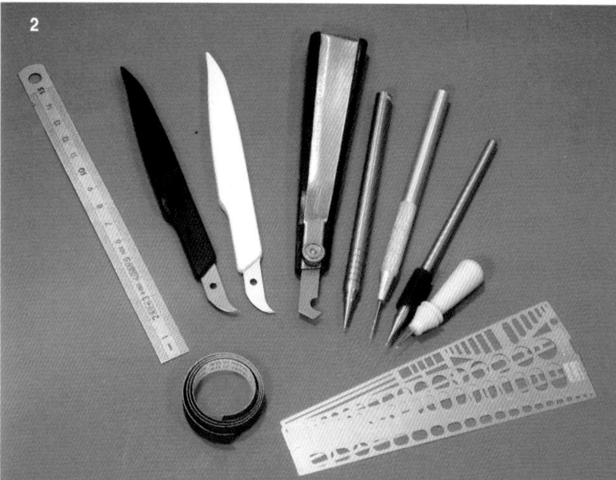

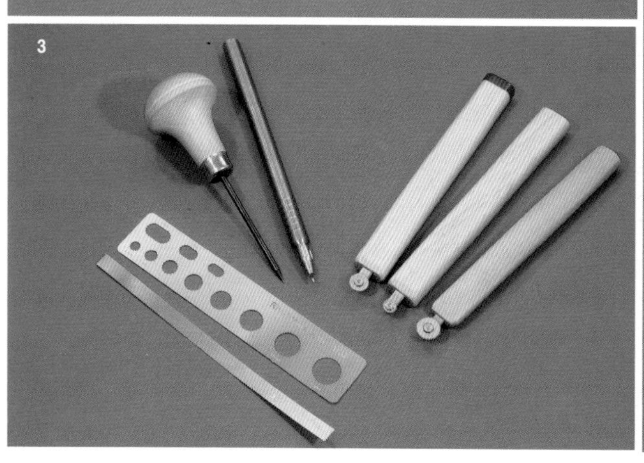

MATERIALS USED

1. The cutters with blades of different shapes that allow easy sculpture or removal.

2. The different tools for re-engraving a kit. The templates are indispensable for re-engraving small panels and can only be used with points.

3. *Rosie the Riveter* is the perfect tool for making rivets on a kit. Easy to use and remarkably precise! The MDC or Hasegawa point allows the making of very realistic flush rivets.

4. All the tools necessary for filing or sanding. The files are used for the bulk of the work. The 400 to 1,200 grain sandpaper is used with water and is useful for any sanding down work. The Micro Mesh is used for the preparation of surfaces and put a shine on the paint.

5. Cements are getting better and better. The very fluid liquid glues join the plastic parts together. The cyanoacrylate cement is also used for the cementing of other materials (resin or metal).

6. The different mastics. The liquid mastics are very efficient for finishing off or for small joints. They can also be used for the making of thicker panels. Double component mastics allow for filling in large spaces or to carry out important modifications.

7. Microset is used for the fixing of the decal whilst Microsol allows it to fit closer to the surfaces (curves, structural lines, protuberances).

8. Paints and thinners. The acrylic paints are diluted with Gunze or Tamiya thinners or with alcohol. Alclad paints, which are lacquer, can be diluted with cellulose or with Tamiya Lacquer. With a brush, Tamiya and Gunze paints dry too quickly. It is better, therefore, to use Prince August paints, which dilute and clean with water.
Enamel paints, like Humbrol, are used just as well with a brush or airbrush.

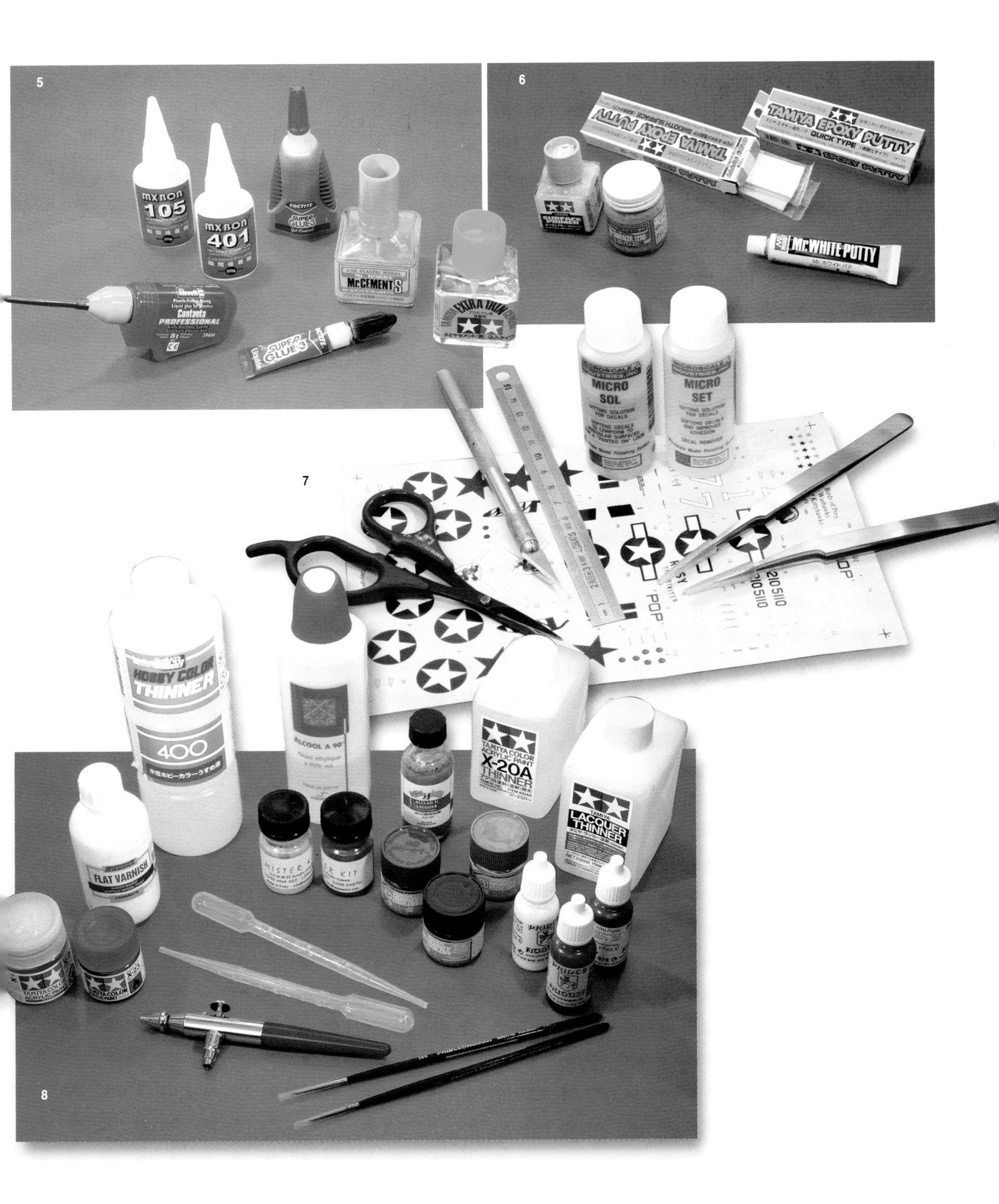
5
mxBON
105
mxBON
401
SUPER GLUE3
Mr.CEMENT S
Contacta
PROFESSIONAL
SUPER GLUE-3
6
TAMIYA EPOXY PUTTY
QUICK TYPE
Mr.WHITE PUTTY
MICRO SOL
MICRO SET
7
HOBBY COLOR
THINNER
400
ALCOOL A 90°
TAMIYA COLOR
ACRYLIC PAINT
X-20A
THINNER
LACQUER
THINNER
FLAT VARNISH
8

CURTISS P-40B
MONOGRAM

1/48 SCALE BUILDING
by Emmanuel PERNES

IT IS DIFFICULT TO imagine the P-40B in anything else but green and brown and with the shark's mouth of general Claire Lee Chennault's AVG, the famous Flying Tigers. It was in these colours that this plane achieved its greatest successes, with many pilots becoming aces, the best known being Robert T. Smith and above all, Greg Boyington, immortalised by the television series *Black Sheep Squadron*. Despite this, the B version is not the best represented of the P-40 by model kit manufacturers.

In 1/48 scale, the best model when it comes to shape and dimensions is still the Monogram, dating from 1964. This means that is a long way from today's standards

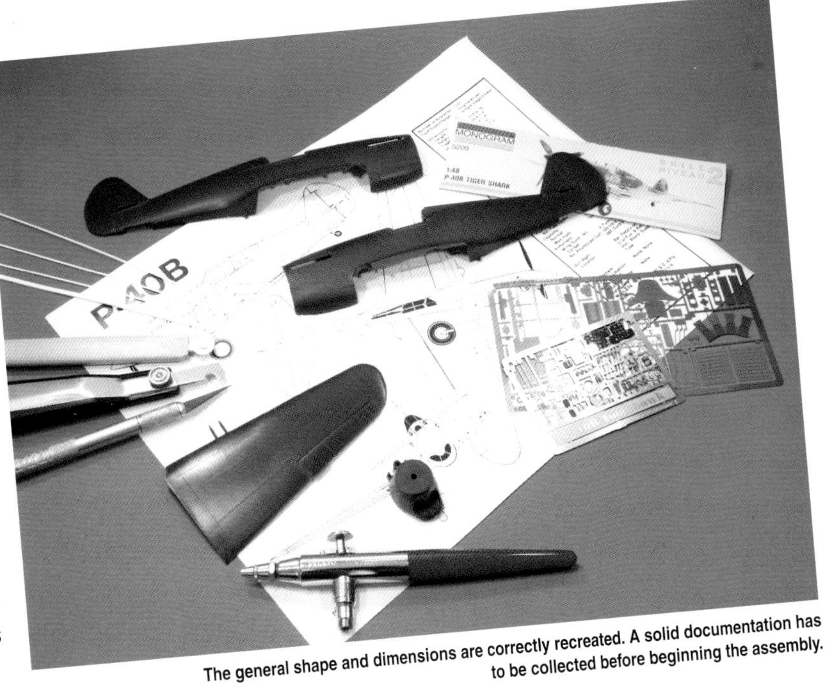

The general shape and dimensions are correctly recreated. A solid documentation has to be collected before beginning the assembly.

when it comes to engraving and interior details. Nevertheless, it is the kit that we have chosen to represent the plane of Tommy C. Haywood, a pilot of the 3rd Pursuit Squadron *Hell's Angels*, based at Kunming, China, in June 1942.

The aim of the following article is to show the tricks of the trade for bringing up to standard an old kit by using several commercially available products (Eduard photo-etching), plastic sheet and the famous 'spare box', where we religiously keep unused parts or the remains of broken kits.

BUILDING PLAN

1 Presentation
- Presentation
- Re-engraving tools

2 Preparing the surfaces
- Engraving and riveting
- Wing fillet option A
- Wing fillet surfacer
- Surfacer
- Ailerons
- Sockets
- Exhaust pipes
- Nose rectifier
- Window rectifier
- Adjusting the added nose.
- hinge

3 Cockpit
- Cockpit preparation
- Scratch cockpit
- Scratch fuel tank
- Painting the fuel tank
- Painting the cockpit

4 Wing surfaces
- Scratch undercarriage wells
- Undercarriage
- Tail wheel
- Hatches
- Flaps
- Scratch flaps
- Anhedral

5 Painting and decals
- Fuel hatches
- Base paint
- Dark earth
- Mask and green
- Painting the upper surfaces
- Wheel masking
- placards
- Decals

6 Weathering
- Wing fillet scratches
- Wash
- Weathering the lower surfaces
- Pigments
- Paint running

7 Finishing touches
- Canopy rail
- Cooling flaps
- Gunsight
- Navigation lights
- Antenna cable
- Thermo shaping
- Finished model

1. Presentation

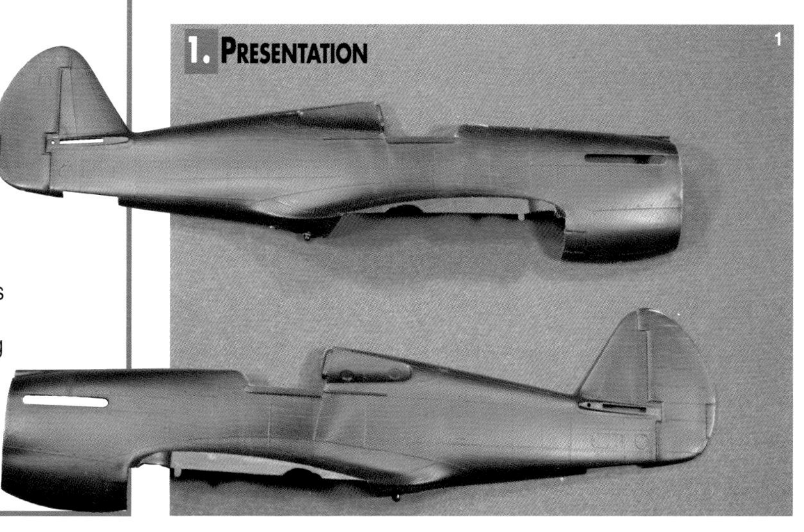

1

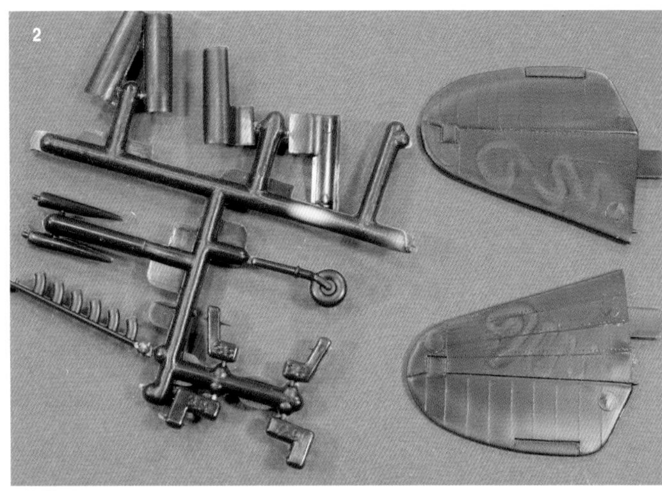

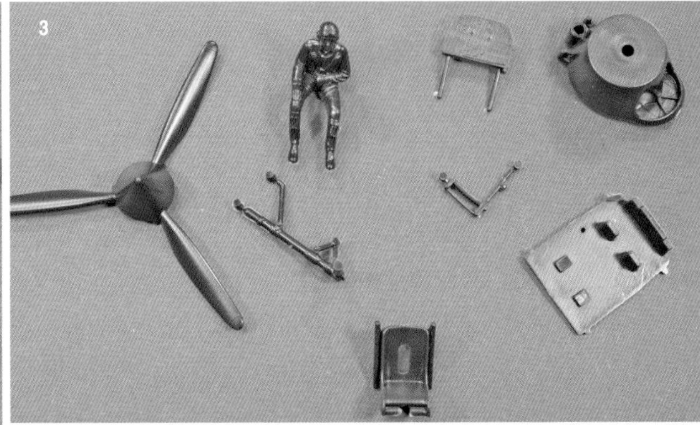

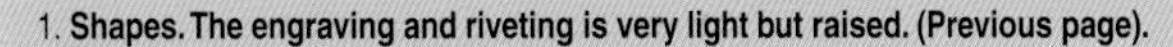

1. Shapes. The engraving and riveting is very light but raised. (Previous page).

2. We see here that the moulding is not always excellent, with a fair amount of smudges and moulding lines on the parts. (here, the elevators).

3. The small parts are often roughly made and often have to be replaced. Some details are extremely simplified, or even false, such as the instrument panel which corresponds to a P-40E!

4. There are several tools available for re-engraving the kit. Templates or Dymo strips will be used to guide. Tamiya of Hasegawa tools (in black) are perfect for straight lines whilst the points are recommended for making curves or small panels.

5. For riveting the surfaces, the best tool comes to us from the Czech Republic. It is a toothed wheel placed on an axis. *Rosie the Riveter* It is extremely precise and very easy to use.

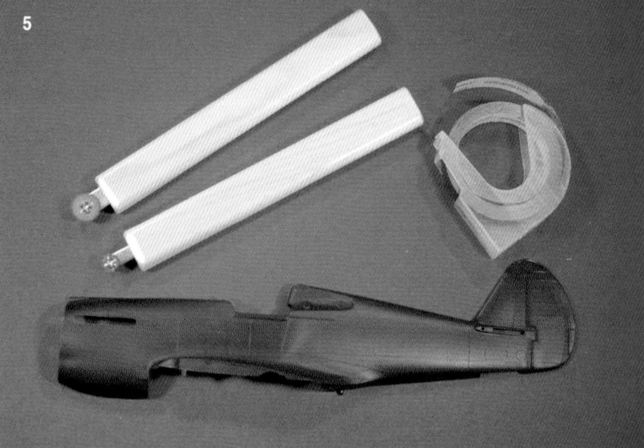

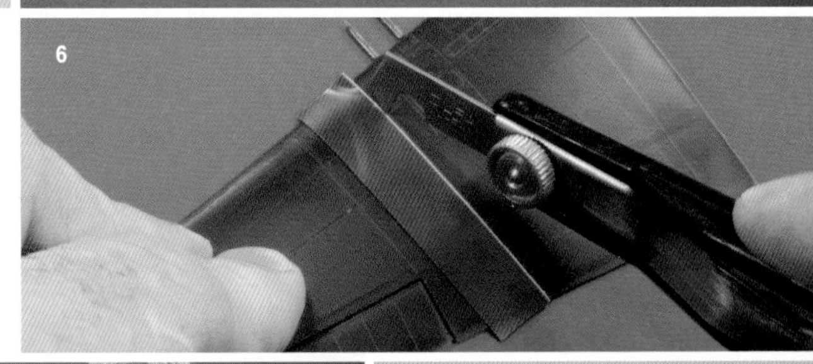

2. Preparation of the surfaces

This is often the most thankless phase, but it really should not be neglected as the plane's final appearance will depend on it. This phase is even, without doubt, more important than the detailing of the internal parts, less visible but more rewarding.

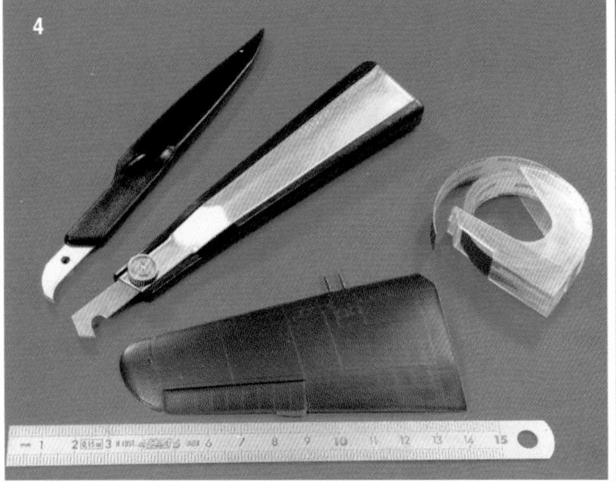

6 and 7. Using the existing engraving (that we will sand down later), we place a strip of Dymo. We next follow this guide to re-engrave by going along it two or three times without pressing too hard. We thus get a sunken structural line by taking out a strip of plastic.

8. We next put a little liquid cement in the line to make it as regular as possible. Once dry, it just needs a little application of very fine sandpaper and water (1,000 or 1,200 grain) to obtain a perfect line.

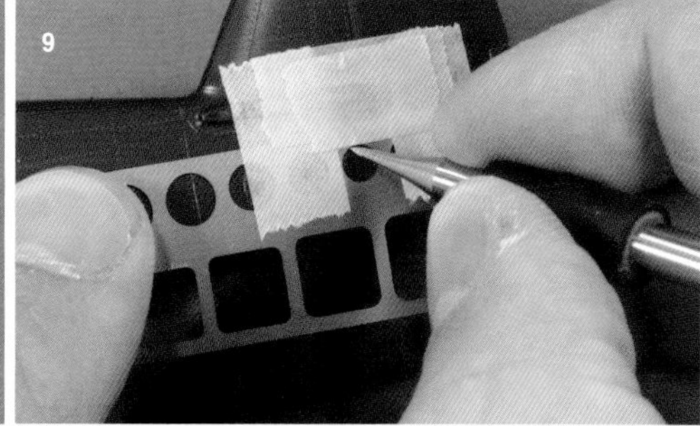

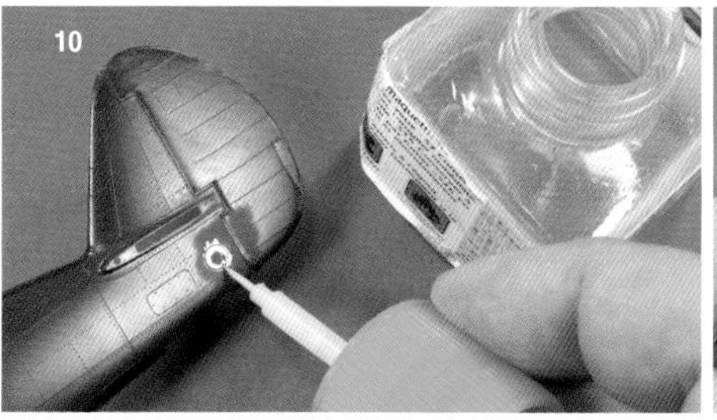

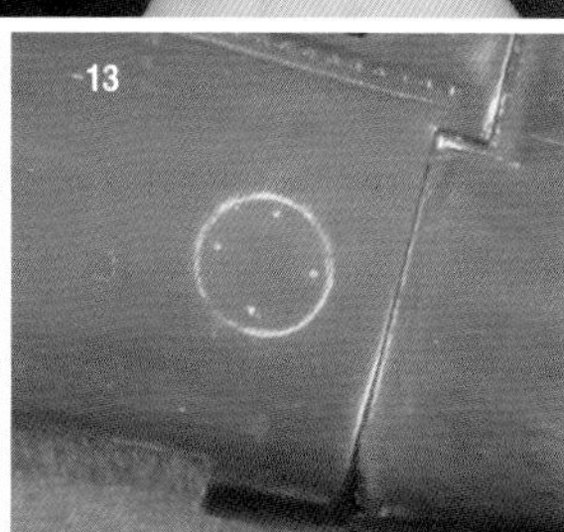

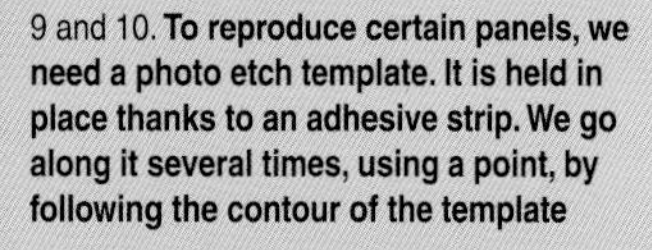

9 and 10. **To reproduce certain panels, we need a photo etch template. It is held in place thanks to an adhesive strip. We go along it several times, using a point, by following the contour of the template**

11. **The result is still uneven. We also have to put liquid cement in the engraving to even out the irregularities.**

12. **Once the cement is completely dry, we gently sand down the engraving.**

13 and 14. **We go over it again without pressing the point in the engraving to finish off this stage.**

15: **The elevators are removed from the tail plane using a photo etched saw so that they can be displayed down.**

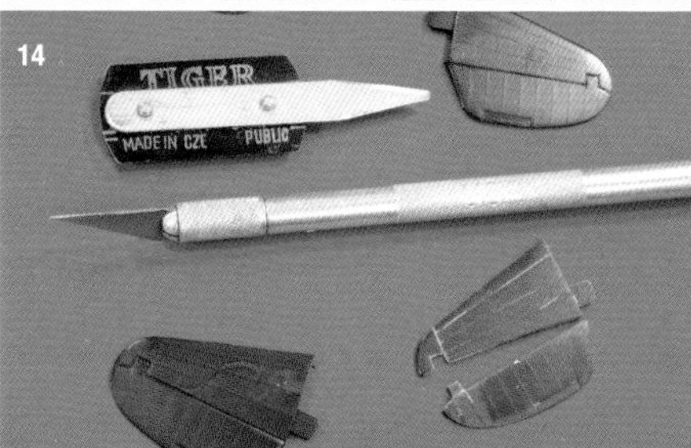

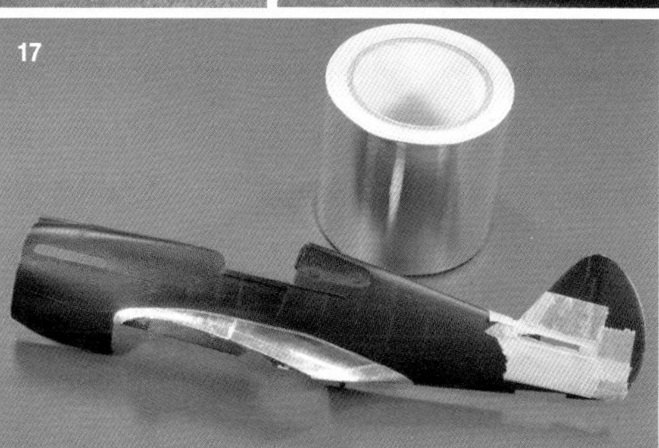

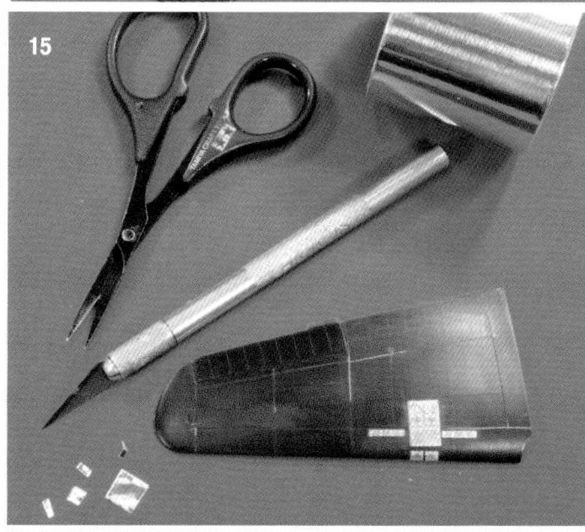

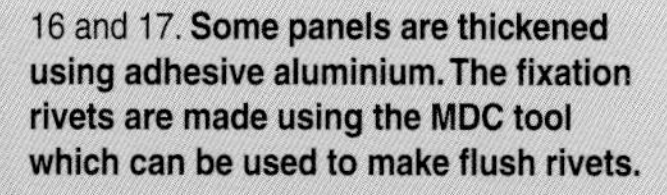

16 and 17. **Some panels are thickened using adhesive aluminium. The fixation rivets are made using the MDC tool which can be used to make flush rivets.**

18, 19 and 20. **Initially, the thickening of the wing fillet was made with adhesive aluminium. However, with much handling, the aluminium scratched and deteriorated.**

Next page
21 and 22. **This is why another solution was envisaged. After precision masking, Surfacer 1000 diluted with Tamiya lacquer thinner, was applied with an airbrush on the areas that have to remain raised.**

23. We thus obtain slightly raised and perfectly smooth surfaces. Also, this product can be riveted and engraved.

24 and 25. The holes for the ejection of cartridges situated under the wings are made. Small pieces of plastic sheet are cemented on to give perfect edges.

26. The exhaust pipes are drilled out. Luckily they are round. We begin by pricking the centre with a point. We then make the holes using bigger and bigger drill bits. The result. We have to do the same again for the eleven others. Luckily they are not fish tails.

27. The air vents made by Monogram are not deep enough and above all, too simplistic. They are firstly taken off with a drill...

28, 29 and 30.... then an X-Acto blade. A rounded file finishes it off.

31. The air vents are lengthened by using tubes cut in two and cemented firmly behind the part.

32. The filters are made from lollypop sticks (of the right size) and the metal grill is cut out with a punch. An optical cover representing the base of the engine block is made to close off the space behind the filters (in case the cooling flaps are displayed open).

33 and 33bis. The final result

34. The added nose does not fit very well, as is also the case with the machine gun cowling. A lot of sanding down, as well as double component mastic are necessary.

35. The lateral parts of the canopy on the left side, that take the fuel tanks on the real plane, are cut out with a drill.

36. The radio access hatch has been slightly thickened with very thin plastic sheet. The piano hinge is a piece of stretched wire striped using an X Acto blade.

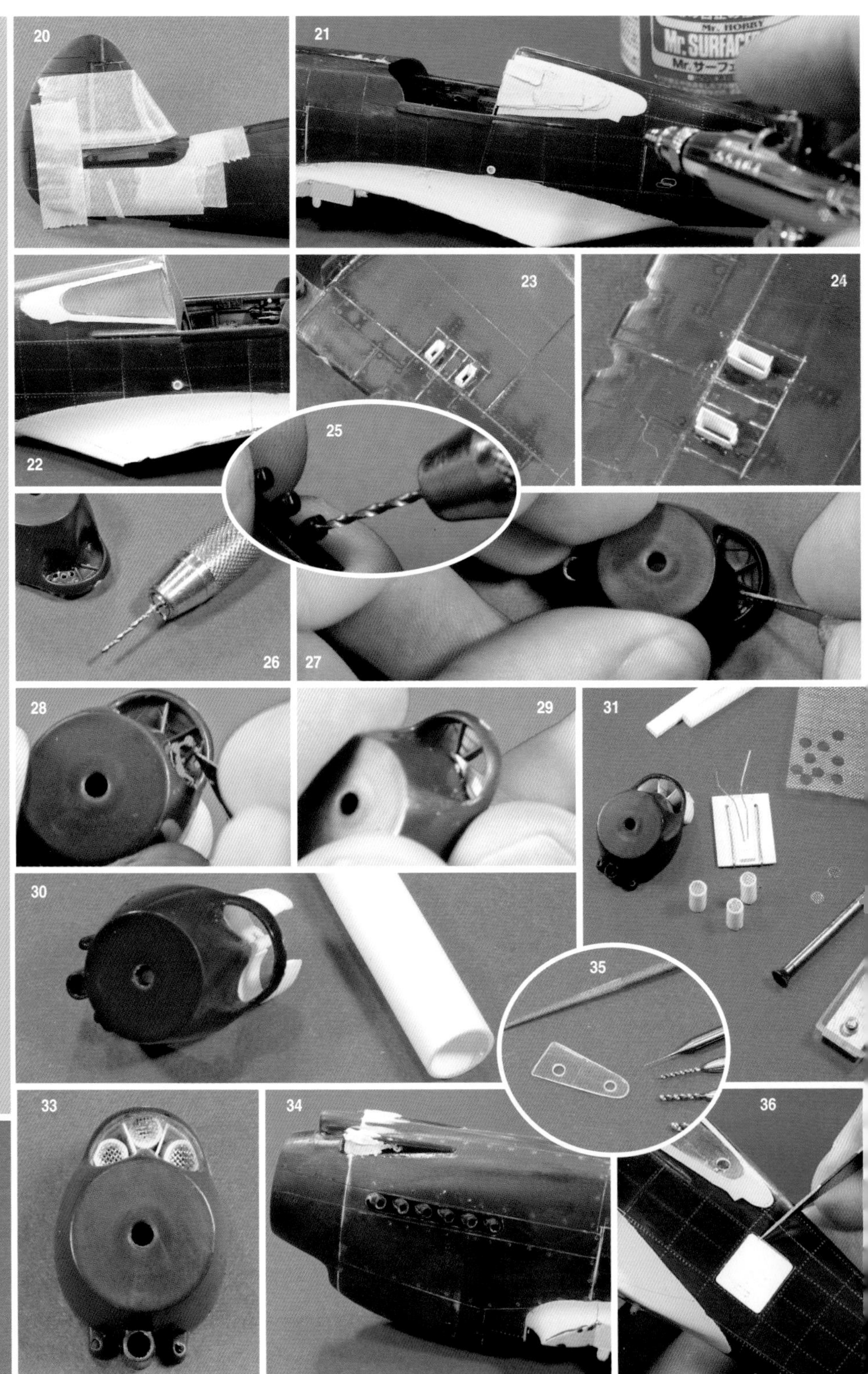

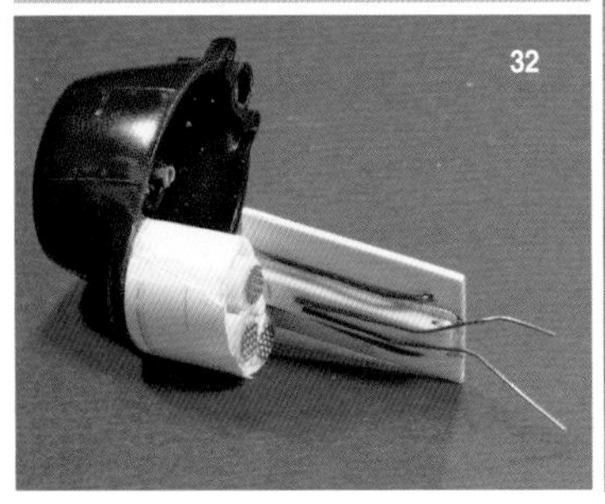

3. The cockpit

The cockpit supplied by Monogram is very basic, even incorrect concerning the floor and instrument panel. It is better to redo it. Eduard has made a photo etch sheet for the Trumpeter P-40 which I used, but plastic sheet will still be useful….

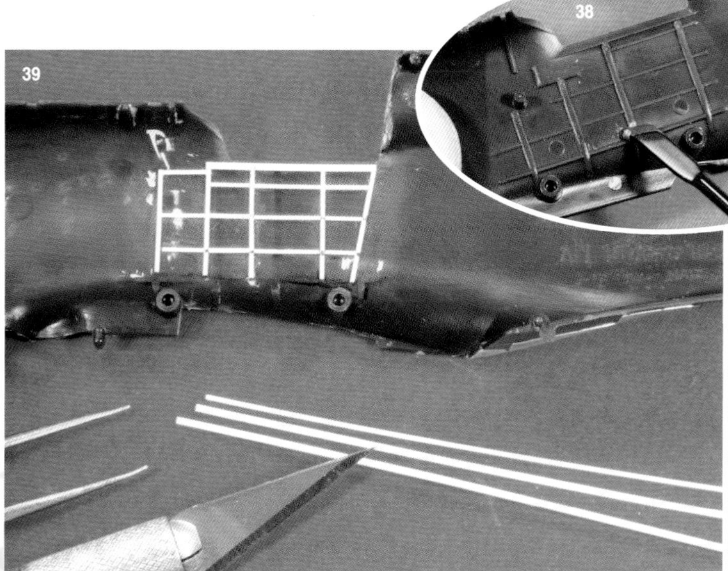

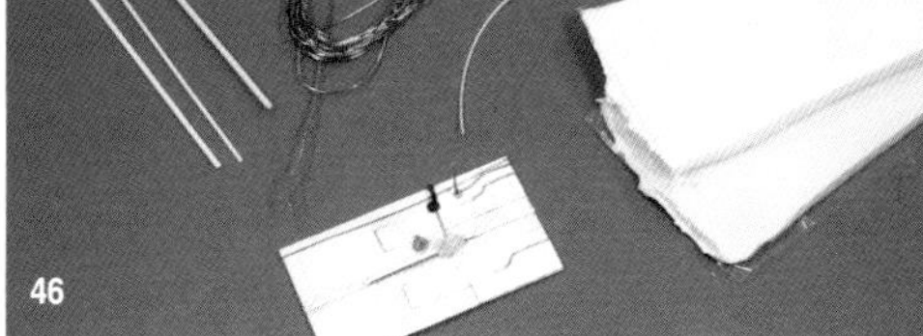

37 and 38. **The engraved structure on the fuselage sides is taken off using a bevelled tipped cutter.**

39. **The sides are dressed using strips of Evergreen.**

40. **The photo etch sheet for the Trumpeter P-40B is very useful. All the boxes are made by folding.**

41. **A small hose is made by winding very fine copper wire around an Evergreen rod.**

42 and 43. **The cables are made with tin wire, which is easily bended and shaped, and fixed with cyanoacrylate cement.**

44 to 46. **The floor is re-made with plastic sheet. It is slightly curved, as with a real aircraft. Small reinforcements are made with plastic sheet and riveted with the help of the MDC tool. The True Details resin seat is given a Tamiya double component resin cushion.**

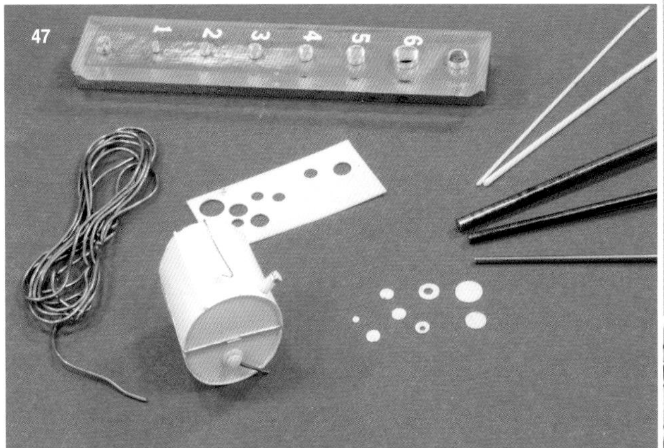

47. **The fuel tank was placed just behind an open partition in the P-40B. Although it is not very visible, it has been scratch built in the usual way and materials.**

48. **The fuel tank is painted in Alclad *aluminium* with *steel* reflections.**

49 and 50. **The patina is achieved with sepia Pébéo acrylic ink. The detailing is copied with Prince August acrylic paints.**

51, 53 and 54. **The cockpit is painted in Gunze interior green H 58. Some lightening is added using Tamiya XF-4 yellow. The usual oil mix of burnt umber reinforces the contrasts. The last stage is the addition of sand and earth colour pigments on the floor, around the areas walked on.**

52. **The pre-painted photo etch adds a real plus to the instrument panel and the harness straps.**

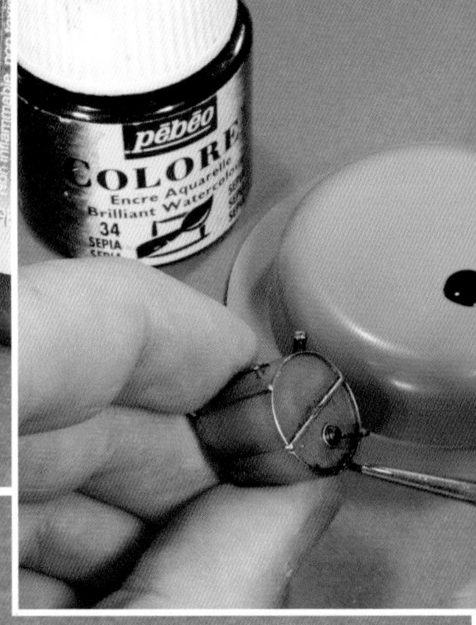

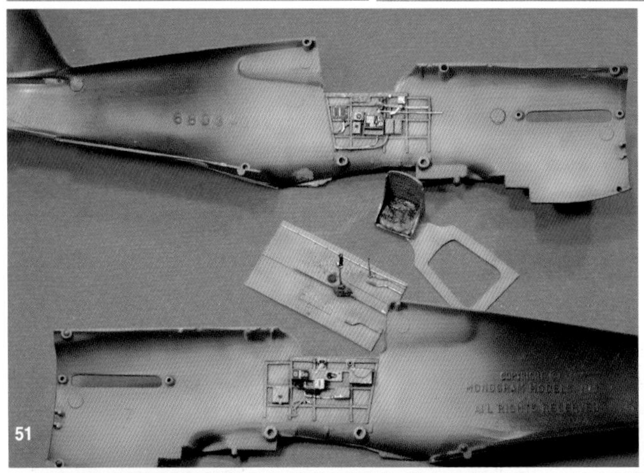

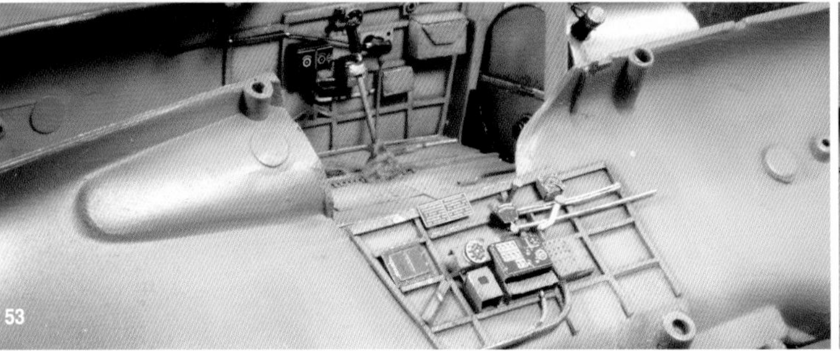

4. Wings

55. The undercarriage wells on the Monogram kit are quite simply inexistent. The first thing to do is partition off the upper part with thin plastic sheet.

56. The area where the wheel is housed is made from several parts. The reinforcements are first cemented…

57.… then the base plate, made separately, is fitted at the end. It goes without saying that there are frequent adjustments before cementing and that, sometimes, badly fitting parts have to be started again.

58. The final result.

59. The undercarriage legs and the wheels as made by different manufacturers, in order Monogram, Hobbycraft, Mauve and Hasegawa. Only having two left hand undercarriage legs in my Monogram spare parts box!, I chose the Hasegawa undercarriage legs. These came from my box of bits, the original model having learnt to fly early.

60. The undercarriage legs have a tin wire brake hose added to them that is 0,23 mm in diameter. The undercarriage doors are taken from the Eduard sheet bought for this kit. They are shaped on the original parts after having been passed through a candle flame to make them easier to shape.

61. In the same way, the Monogram supplied flaps have been replaced by the much thinner photo etch Eduard flaps.

62. The insert supplied by Eduard is too big for the Monogram kit. However, we can use the parts to make a new one in plastic sheet.

63. Some parts are difficult to reproduce using scratch building (perforate panels, reinforcements) and were taken on the photoetched fret.

64. During the 'crucifixion', the correct wing angle is held in place with Tamiya adhesive tape when cementing.

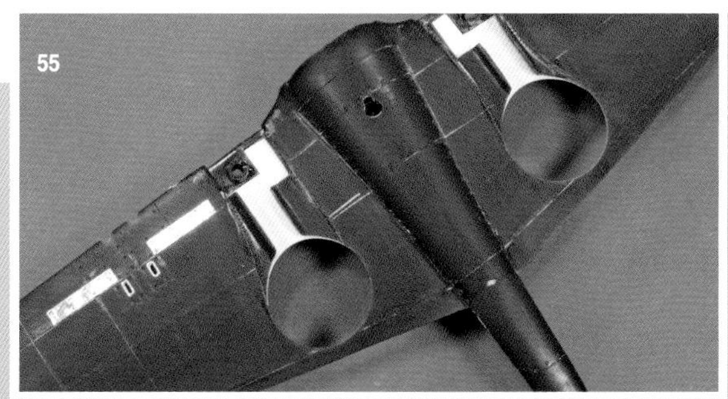

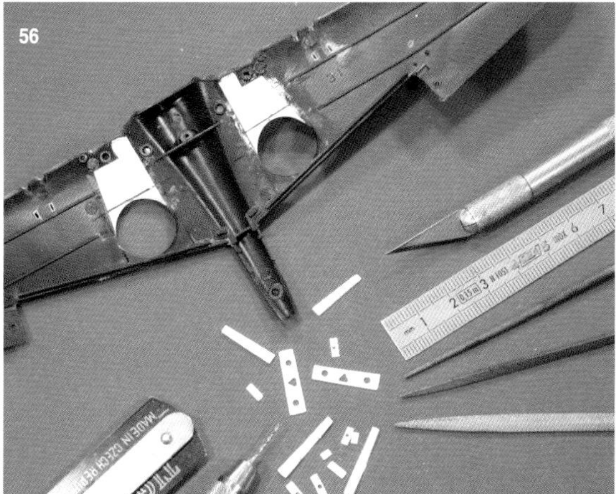

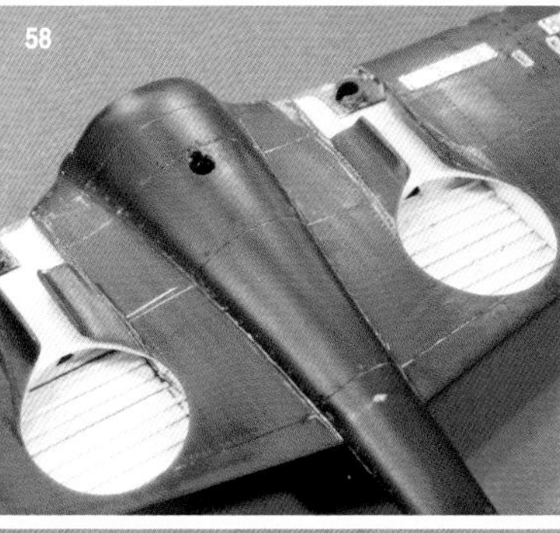

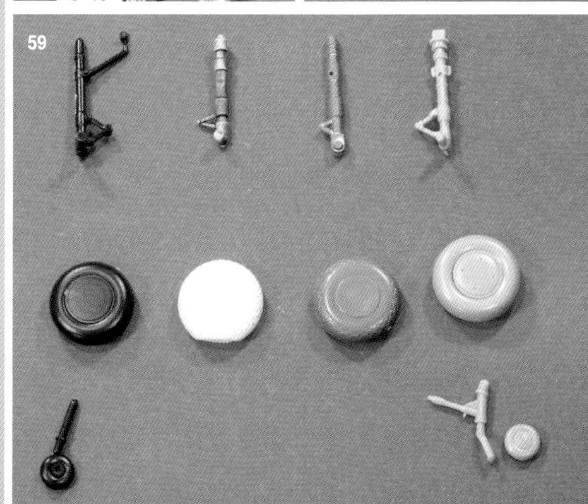

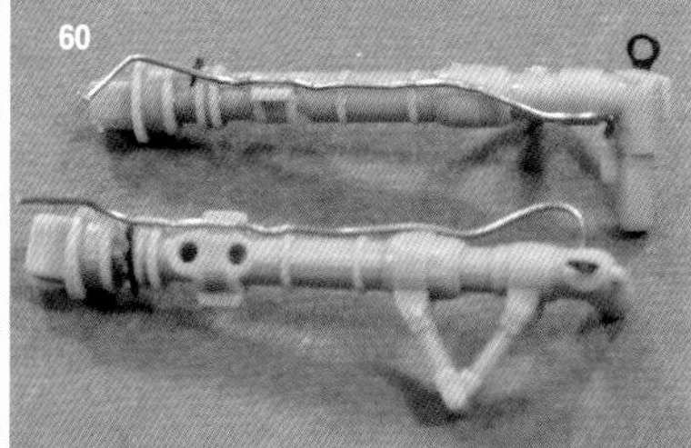

61

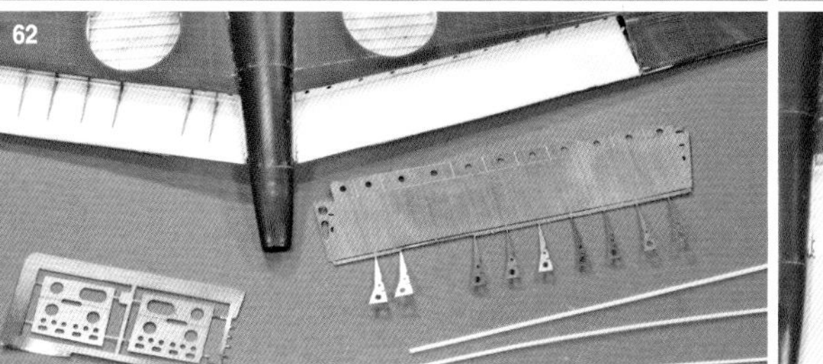

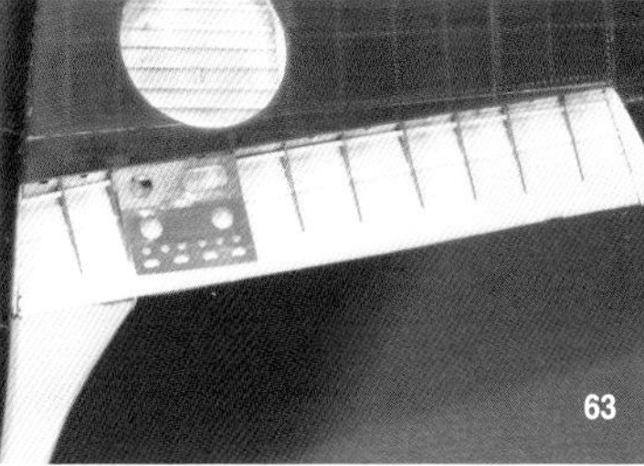

5. Painting and decals

Tommy C. Haywood's aircraft, which has been chosen to illustrate this article is really colourful with a beautiful shark's mouth, several flying tigers and personal insignia, a bulldog's head with horns.

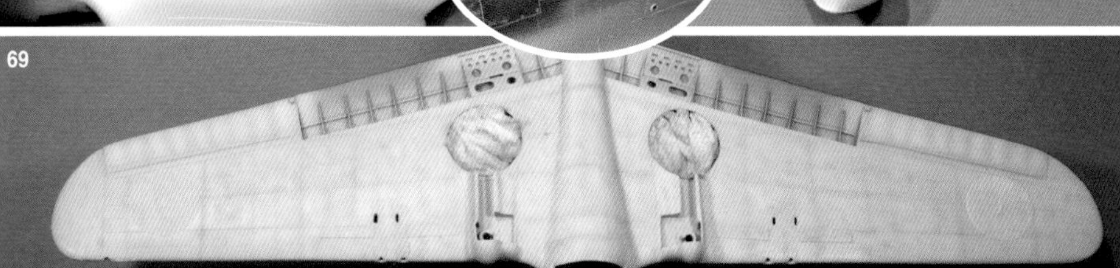

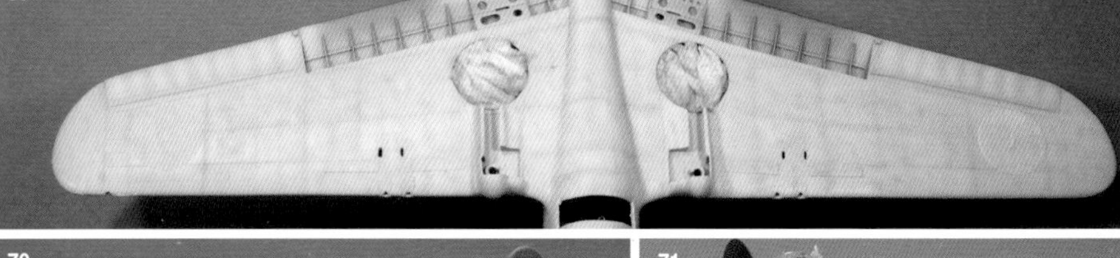

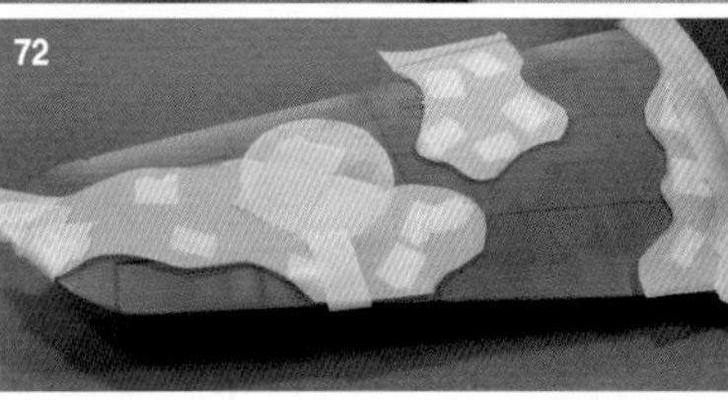

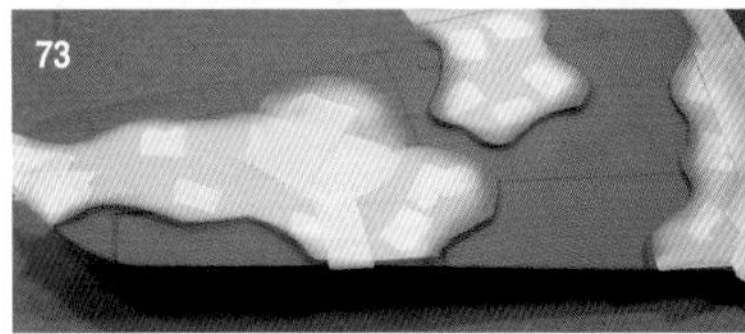

65. **The parts in front of the lateral canopy areas are not green, but the colour of the camouflage. They are, of course, painted and weathered before the canopy glass is cemented.**

66, 67 and 68. **Before adding the camouflage to the areas worn through maintenance, an aluminium base paint is applied. By masking them before painting, or by scratching after, we can obtain good results of wear and tear. Note that the windshield has only now been cemented and the mastic applied (I should have done this before!)**

69. **The upper areas are gloss grey H 61 Gunze. The round masking is placed in the area originally reserved by the manufacturer for the British roundels. The grey is re-worked with white and black to give the impression of weathering. The surfaces are *whipped* in the relative wind direction and *mottled* to make it look more natural.**

70 and 71. **The AVG planes, originally destined for the United Kingdom, are given a British type camouflage of dark earth and dark green. However, in reality, the planes were painted in the United States in approximate shades made by Dupont; they differed slightly from those used by the RAF. Dark earth is more ochre and it is, therefore, Gunze H 304 that has been chosen to imitate it.**

72 and 73. **The Dupont dark green was similar to the American medium green 42, here Gunze H 302. The delimitation between the two shades on the upper wing surfaces is quite clear as they were factory painted using floating masking. It is this method that has been used. The camouflage is copied with tracing paper on a 1/48 scale blueprint and small pieces of rolled adhesive tape keep them in place..**

74. **Care must be taken to keep the airbrush at the perpendicular to the masking to avoid paint going under the masking and spoiling the final result.**

75. **The final result is realistic. The delimitation is quite clear. The paint of the upper wing surfaces is treated in the same way as the lower wing surfaces. It is better to avoid the white and black to lighten or darken the green or brown (this could result in a chalky effect) and to use instead, yellows, sand or dark browns.**

76. **An efficient way of painting the wheels is to start with the tire, then spread, by capillary action, liquid masking around the hub.**

77. **Thick coats of fresh paint, therefore without weathering, where insignia is to be placed, are added after painting.**

78. **Between two coats of gloss varnish (Tamiya X 22), the decals from the Cutting Edge sheet are applied with the help of Microscale softening products.**

79. **The flying tiger under the cockpit came off in flight, rendering visible the varnish that had been painted over it to protect it. This detail has been reproduced by Cutting Edge.**

80. **A new flying tiger, placed in the V for Victory, featured on both sides of the fuselage. The roundels were of a strong blue originally, but faded quickly, another detail well reproduced by the American brand.**

6. Weathering

81 and 82. **The walk areas of the wing fillets are very scratched on the 'Flying Tigers' P-40 planes. This effect is easy to achieve. After the aluminium base paint, we dip a piece of foam in liquid masking then dab it along the chosen area. After painting the camouflage, we rub the masking off and the aluminium re-appears.**

83. **To bring out the structural lines, we use a brush to apply a very diluted oil wash. Any excess wash is removed with a soft rag.**

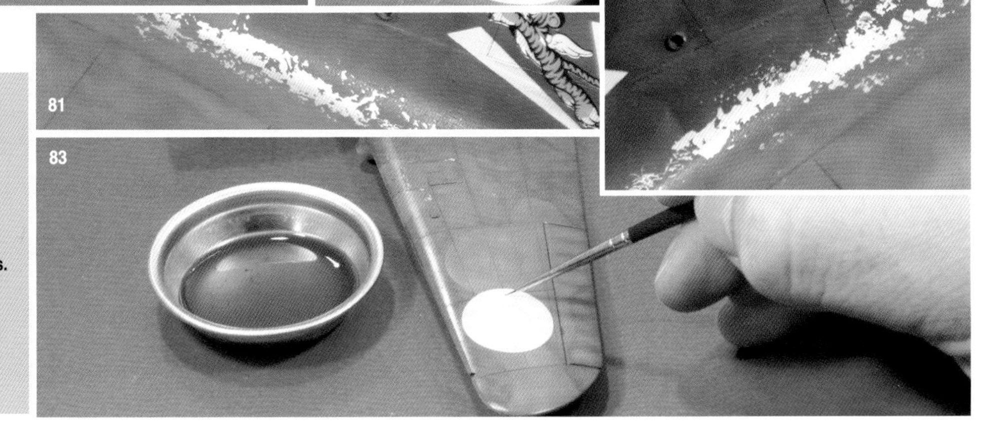

84 and 85. **Certain structural lines are highlighted using the airbrush with a very diluted (90% diluted) mix of brown and black thanks to a piece of Post it paper. Oil drips are achieved with acrylic sepia Pébéo ink mixed with matt varnish and applied with a thin brush.**

86. **The walk areas at the wing roots, as well as the undercarriage wells have CMK or Tamiya earth pigments added to simulate the dust or mud that gathers in these areas.**

87. **Fueldrips from the fuel tank caps are achieved thanks to Pébéo inks and burnt umber artist's oil paints. The exhaust smoke stains are airbrushed with cream (Gunze H 318) and highlighted with a thin stroke of dark brown (still using the airbrush).**

7. Finishing touches

88. **The canopy rails, left out by Monogram, are made with *U* shaped Evergreen strips, cut in two and adjusted with Tamiya ultra fluid liquid cement.**

89. **The cooling flaps, originally part of the fuselage, come from the Eduard sheet and are folded into shape. They add a realistic touch to the model.**

90. **The cruciform sights and cross hairs are very fragile and added right at the very end of assembly. The machine gun barrels are Quickboots 12,7mm in 1/72 (to simulate the 7.7 mm in 1/48)**

91. **The teardrop shaped navigation lights, characteristic of the P-40 are taken from an Aires sheet.**

92 and 93. **After having made a hole with a very thin drill, we insert the light and fix it with white cement. The result is very pleasing.**

94. **Placing the antenna wire on a P-40 can seem very complicated. The first stage consists of making a small copper hook that we cement to the top of the tail fin with cyanoacrylate in a previously made hole.**

95. **After having drilled a hole next to each position light, we cement with cyanoacrylate, a piece of 0.065 mm fishing line to one of the two sides.**

96. **We thread the line through the hook and cement it to the other wing. It does not matter if the line is not taut. To make it taut, you just have to pass something hot two centimetres underneath (the glowing end of a matchstick, heated X-Acto blade….) the line and it will go taut. You have to be careful, of course, otherwise the line will break; the result, however, is good.**

97. **The last piece of line is easier to cement and attaches to a small insulator placed on top of the fuselage.**

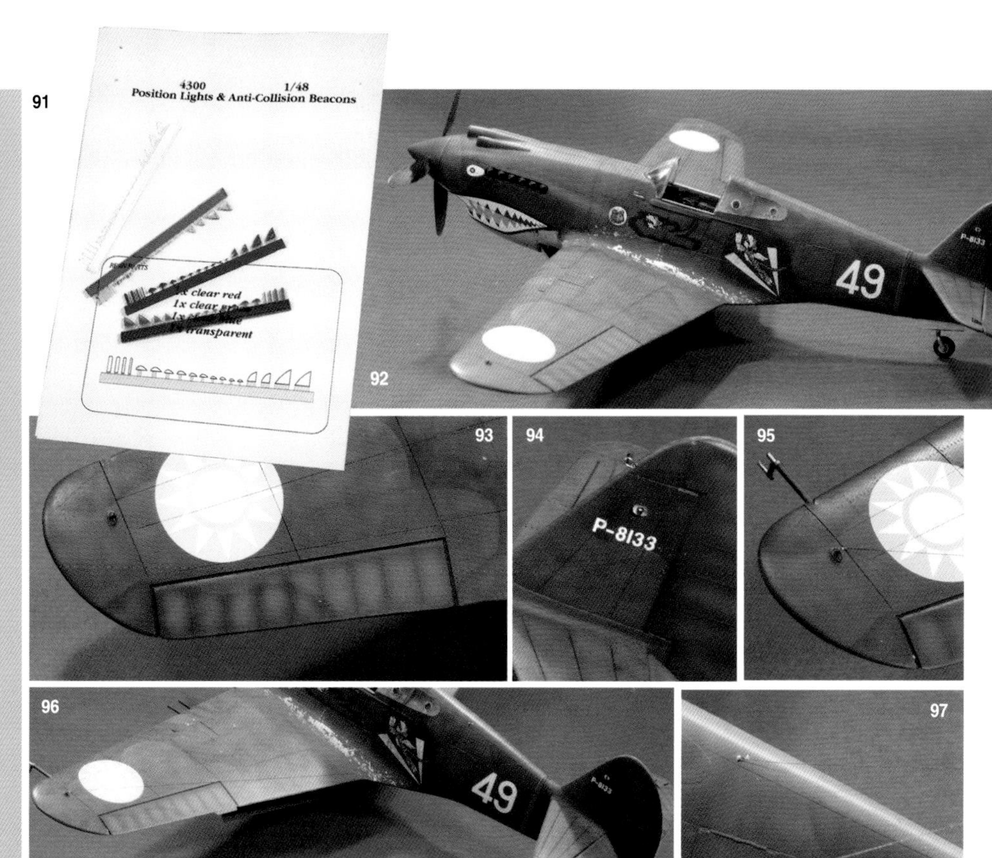

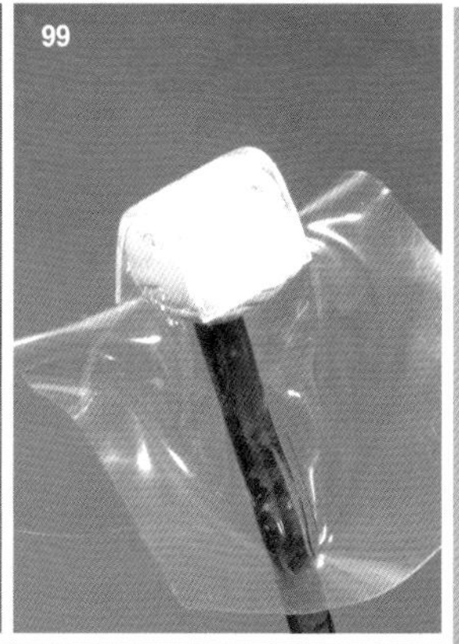

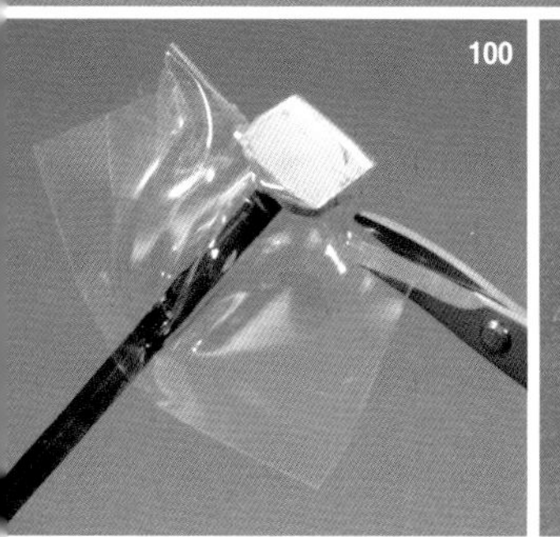

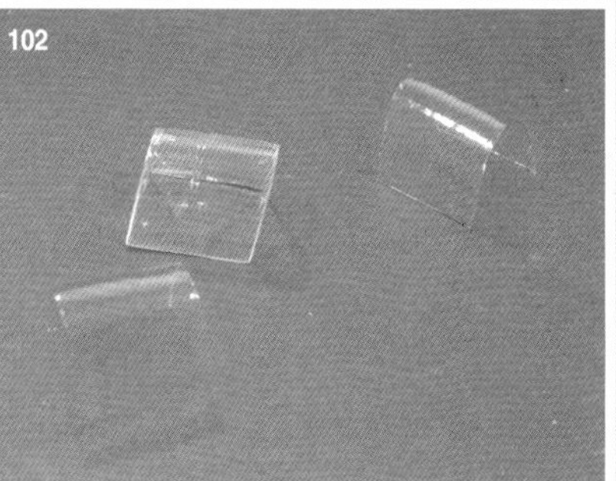

98. **The last stage of assembly consists of thermo shaping the sliding part of the canopy, too thick originally for it to remain open. The necessary material is basic for a simple shape, a piece of transparent plastic, the original part, a stick and a candle.**

99. **The canopy is held by Blue Tack on a stick held in a vice. The transparent plastic is held over the flame until it becomes shiny. Finally, we place the plastic on the canopy by pulling it down and by bringing the sides towards the interior.**

100. **Begin cutting out the new canopy with scissors to start with. We finish off with a new blade and by sanding, using the original canopy as a stencil.**

101 and 102. **The canopy is soaked in Klir to strengthen it and give it a good transparency.**

Next page. **Finished model. Many illustrators show this aircraft's tiger on a thick dark green layer. In reality, it would appear that this was well protected by gloss varnish which made this area look darker in photos.**

49
P-8133

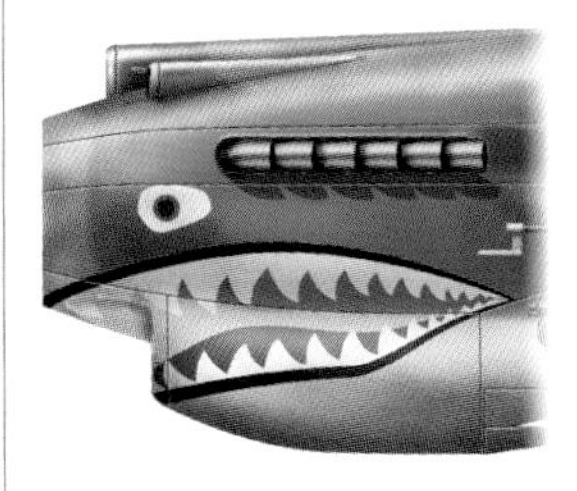

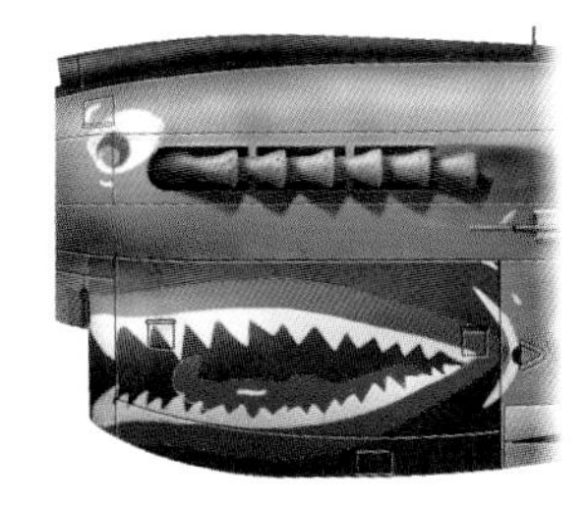

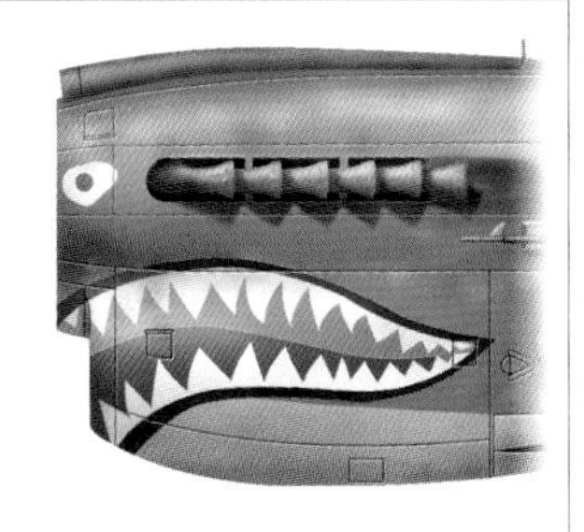

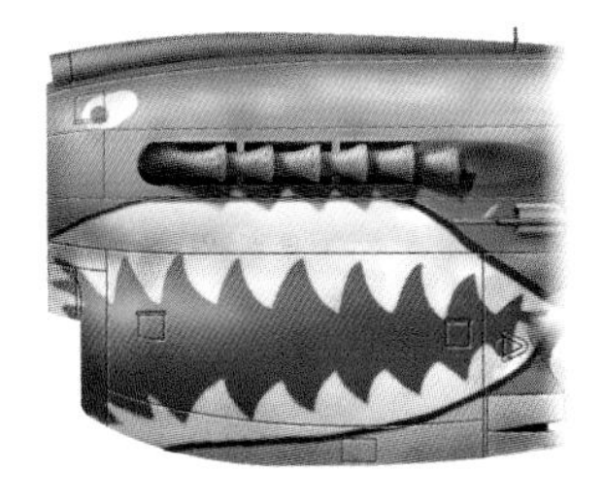

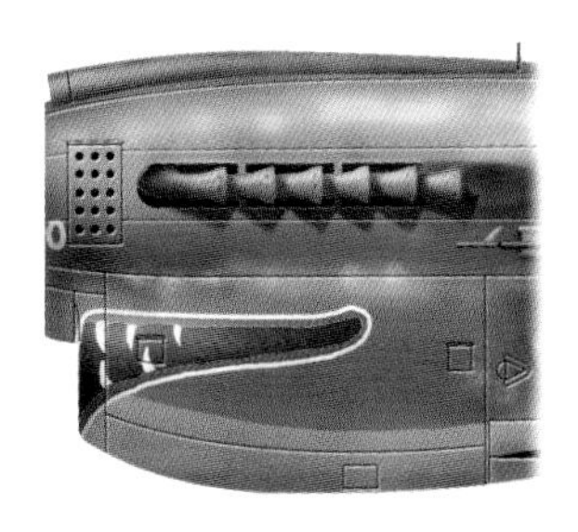

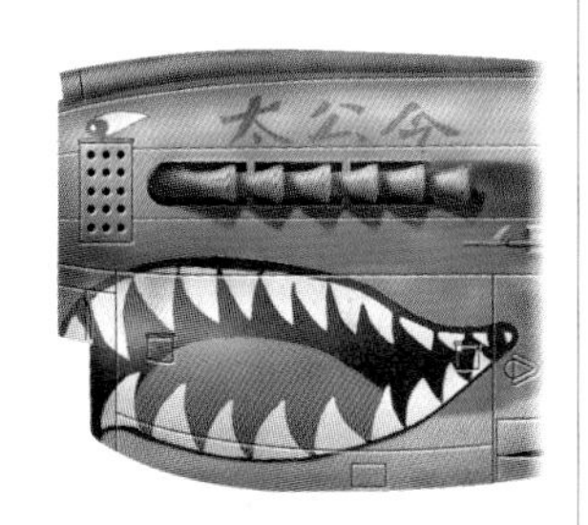

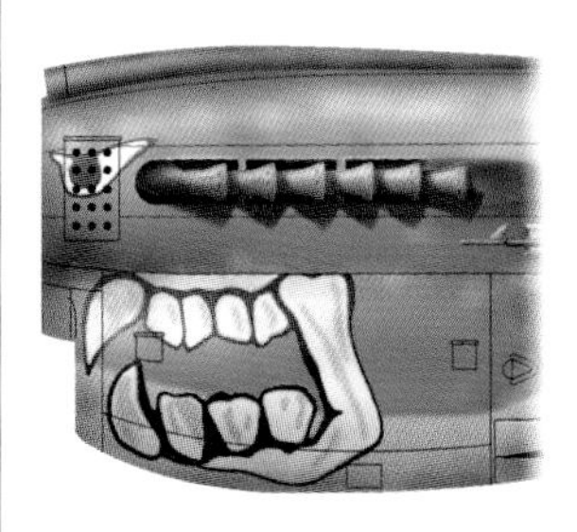

'SHARKMOUTH'

One of the principal features of the P-40 was the 'Sharkmouth'painted on the air intake and made famous by the 'Flying Tigers'. Here is a selection of those found on almost every version of the fighter during the Second World War. Most of the Skark's Teeth variants appear in the book '*Curtiss P-40, from 1939 to 1945*'(Planes & Pilots series n°3) and may be consulted in order to check on the unit and the version. The only exception is the white skull, painted on the front of the P-40Ns of the 80th Fighter Group in India in 1943-44. There is an original specimen on the far left of the second last row which shows only the jaw after the top had been taken off when a panel was changed. *(Drawings by D. Laurelut © Histoire & Collections)*

THE P-40E WAS, ALONG WITH THE P-40N, the most produced version of the Curtiss fighter. It saw action in all the theatres of operations of the Second World War, from the Aleutian Islands to Australia, Russia or Africa. The aircraft that we have chosen was part of the 9th Fighter Squadron of the 49th Fighter Group that, in 1942, protected the town of Darwin in Australia from Japanese air raids. Its exterior was plain, classic, but not without originality, and will bring out the modification work that falls within the scope of this article. The last few years have seen an abundance of extra detailing products for model kits. They are, in general, made of resin, and are accompanied by more and more beautiful and precise photo etch parts. The leading brands in resin parts are CMK and Aires, whereas Eduard dominates the photo etch market. We have chosen their products to show you the different stages of advanced detailing of the 1/48 scale Hasegawa Curtiss P-40E kit.

1. Preparing the parts

1. **Hasegawa has got into the habit of sometimes excessively cutting out its kits in order to sell as many different versions as possible.**

2. **The internal details are excellent and will be only slightly improved by the addition of pre-painted Eduard photo etch parts.**

3. **Some details are delicately removed from the side control panels with Hasegawa scissors. They will be replaced with photo etched parts.**

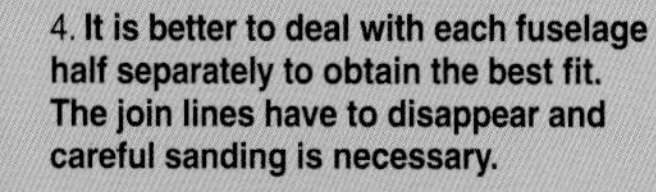

4. **It is better to deal with each fuselage half separately to obtain the best fit. The join lines have to disappear and careful sanding is necessary.**

5. **The undercarriage wells are basically detailed with Eduard parts...**

6. **.... as are the side control panels.**

7 and 8. **To obtain the particular design of the P-40 seat, we use a propelling pencil.**

To detail the aircraft, the Squadron published Walk-Around on the P-40 will be of great help.

PRESENTATION

1. Preparation
- Cutting out the engine
- Cutting out the armament
- Riveting

2.The cockpit
- Painting the cockpit
- Instrument panel
- Cockpit weathering
- Finished cockpit

3. Assembly
- Wing fillet improvement.
- Undercarriage
- Lateral glass sections
- Scratch oil tank
- Scratch fire wall
- Painting and weathering
- Decorations
- Painting the armament
- Masking the undercarriage wells
- Base paint
- Painting the lower surfaces
- Painting the upper surfaces

5. Decals
- Propeller decals
- Matt varnish
- Lower surfaces weathering
- Exhaust marks
- Finished airframe
- Undercarriage details

6. The engine
- Engine compartment
- Painting the engine compartment
- Engine presentation
- Resin preparation Engine
- Painting the engine
- Engine shaft
- Scratch exhaust
- Filters
- Painting the exhaust
- Finished model
- Preparing the parts

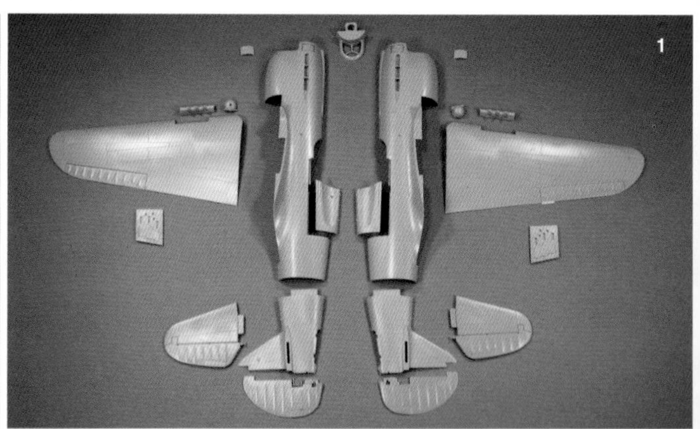

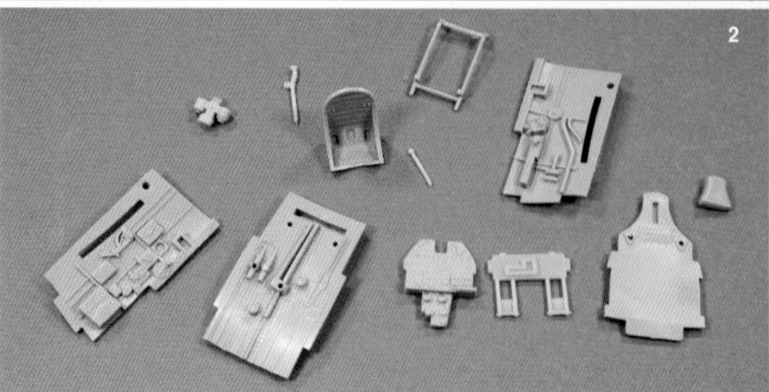

CURTISS P-40E
HASEGAWA
SUPER DETAILING A KIT

1/48 SCALE BUILDING

by Emmanuel PERNES

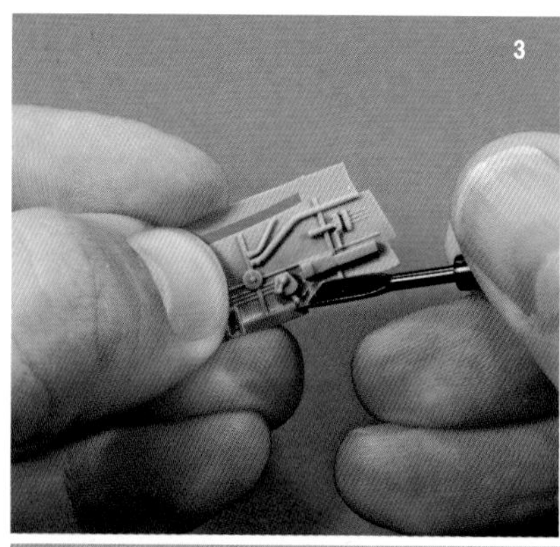

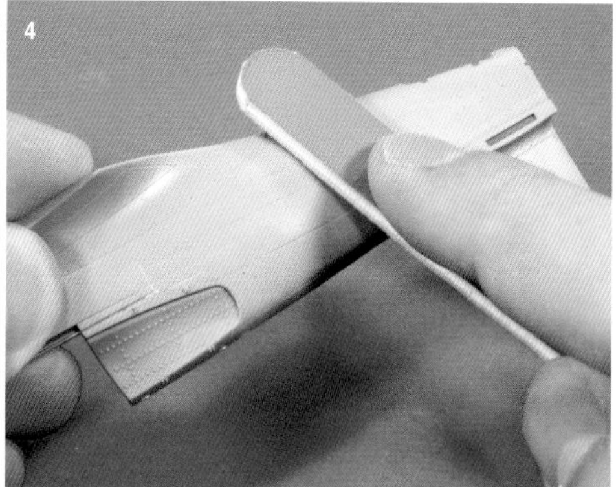

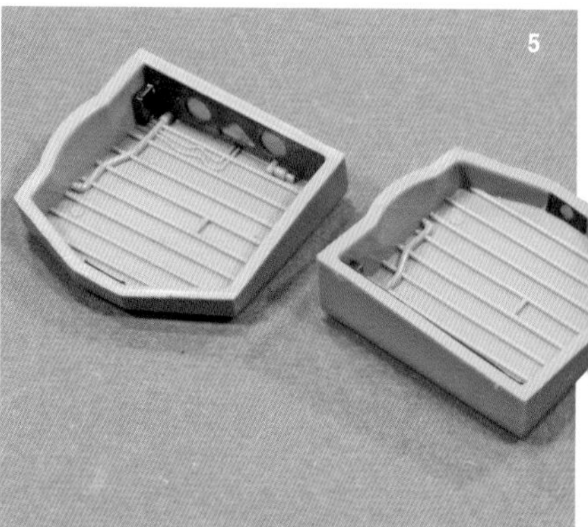

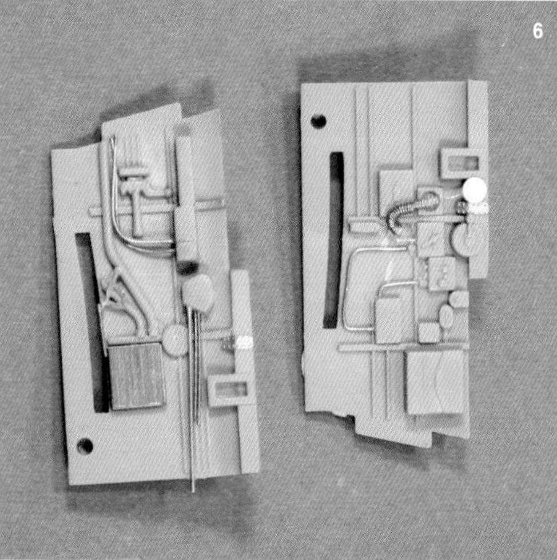

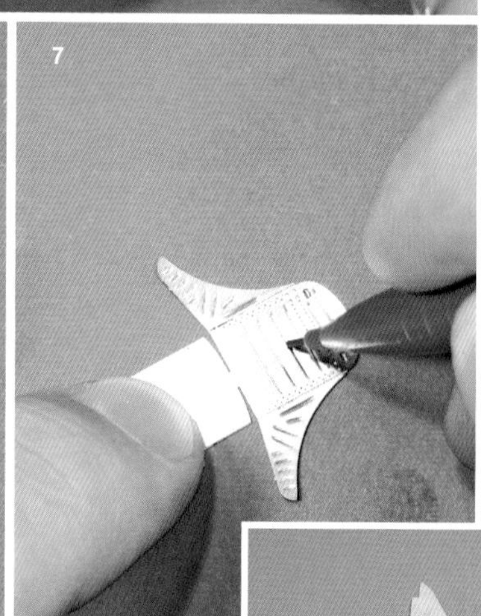

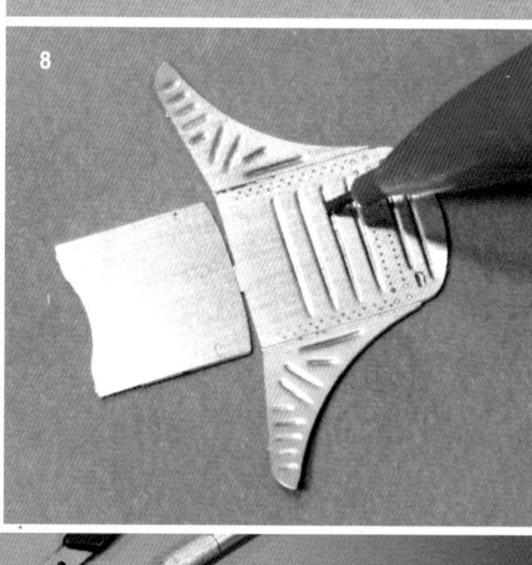

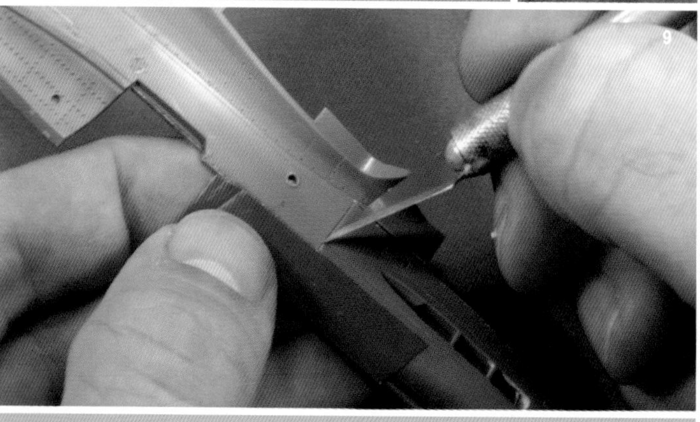

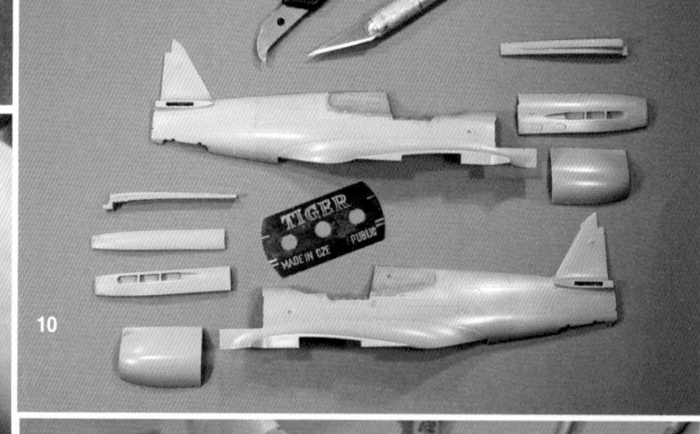

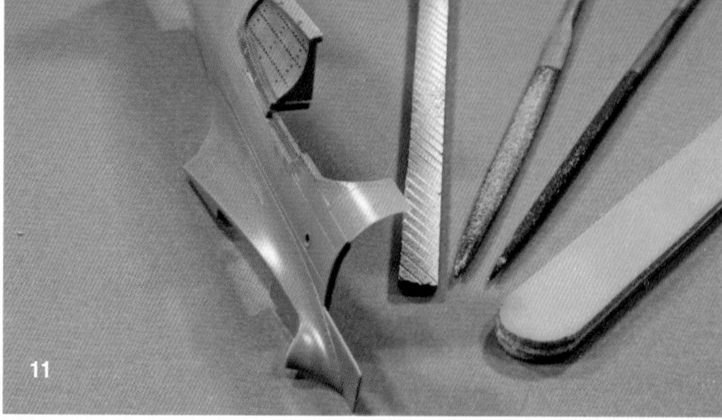

9. To show the engine, the engine hood has to be cut out.
As these parts cannot be found in a shop, great care must be taken in cutting them out so that they can be used later.

10. Cutting out starts with an X-Acto blade or an engraving point and is finished off with a Tiger photo etch saw.

11. The edges are thinned down to respect the scale.

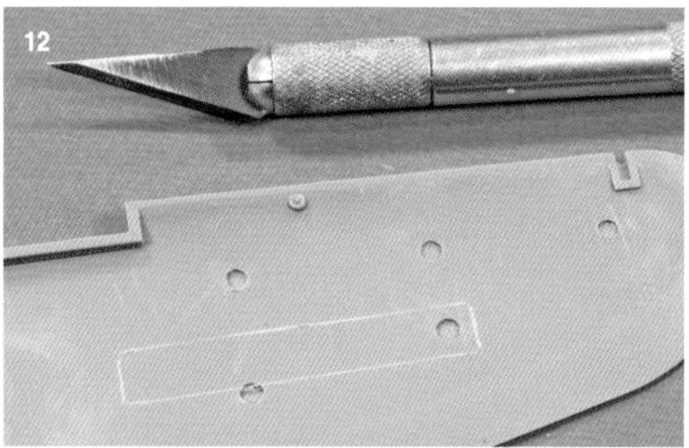

12 and 13. **The same cutting out technique is used for the armament panels. When the line made with the X-Acto blade starts to appear at the rear of the part, we can begin to take out the part we need to remove.**

14. **The Acto-X blade is used to perfect the cutting out, followed by files.**

15. **We use a curved blade to thin down the other side of the part so that we can insert resin parts.**

16. **The piece on which the resin part is moulded is cut using a saw.**

17. **The CMK resin armament compartments are very precise and fit with absolutely no problem. Dry trials are necessary to make sure though!**

18. **The lines of rivets are first drawn onto the surfaces, using a plan.**

19. **We follow these lines with the toothed wheel *Rosie the riveter*. The tool is easy to use and does not need a guide.**

20. **The result that will allow for several weathering effects after the general painting.**

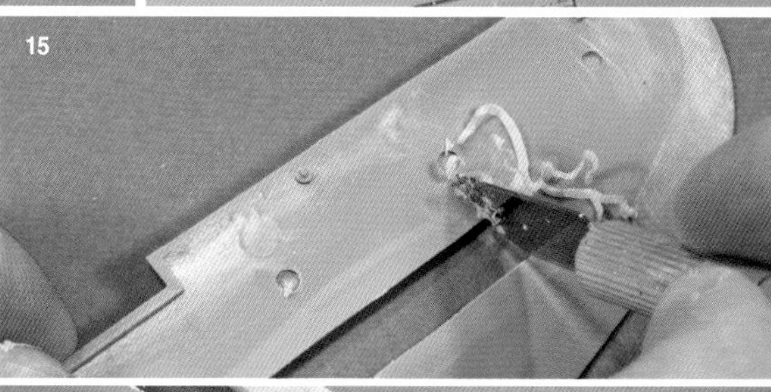

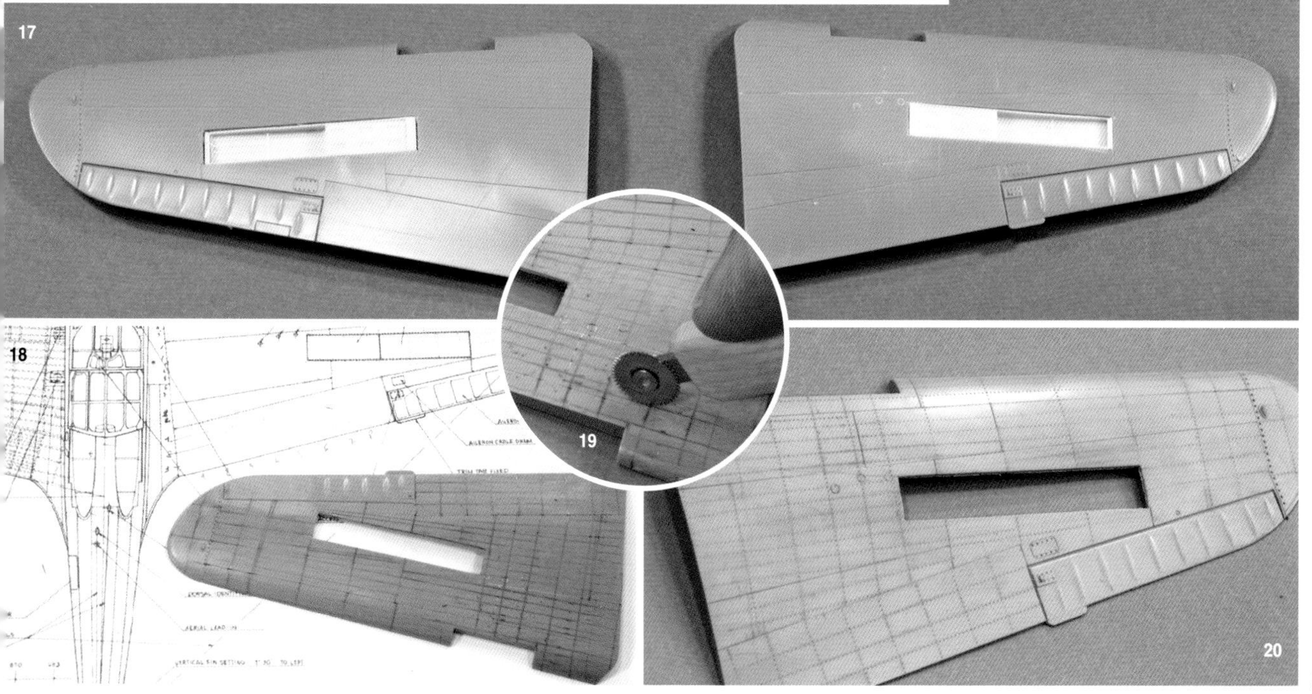

2. The cockpit

The Hasegawa supplied cockpit is excellent. The floor, notably, is superb and we can, therefore, make do with a good paint for it to be convincing. However, the pre-painted photo etch parts, for some of them at least, such as the seat, the straps or the instrument panel, are an undeniable plus.

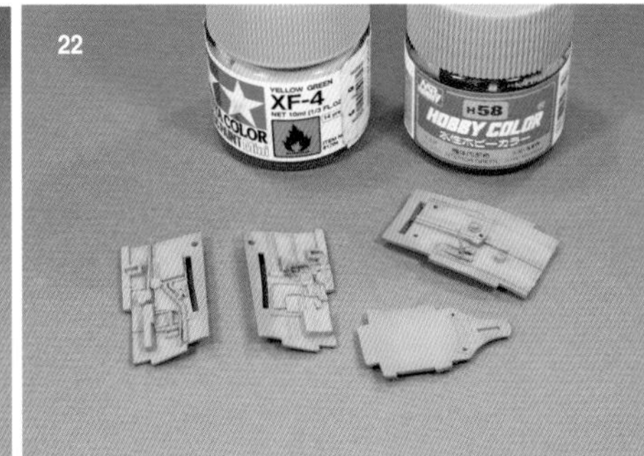

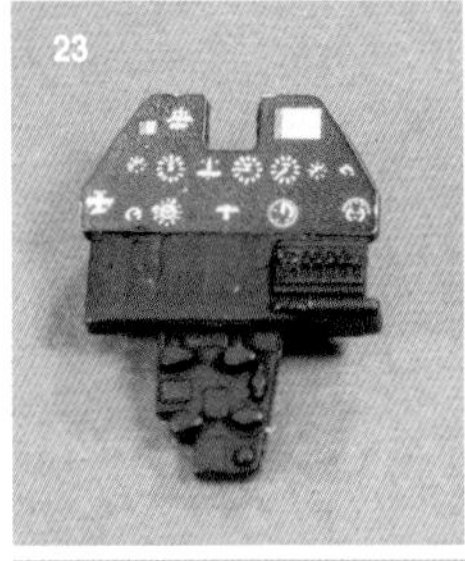

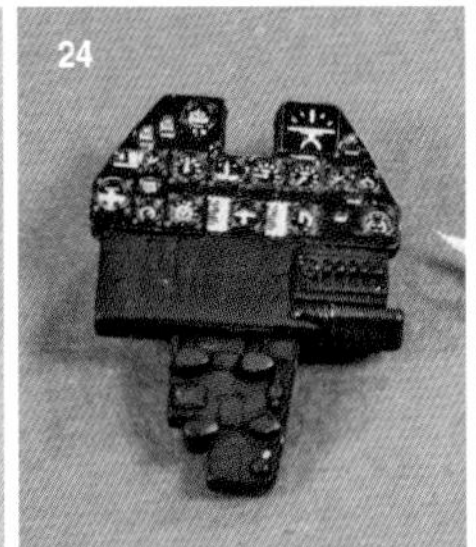

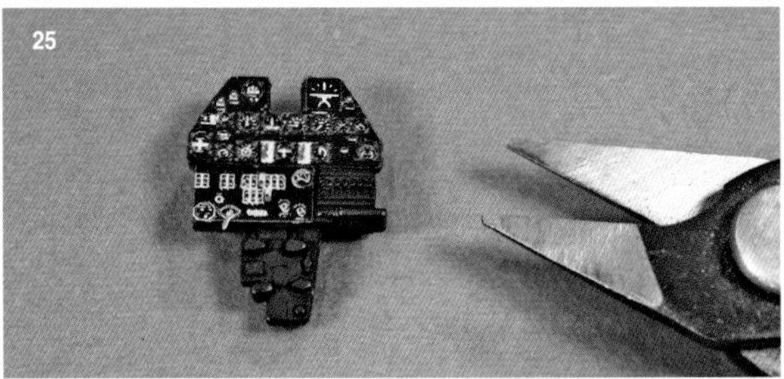

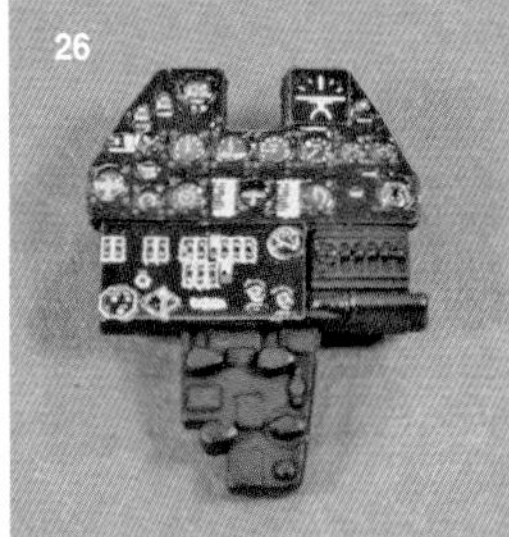

21. The cockpit is painted in Interior Green (Gunze H 58)

22. The colour tone is re-worked with green yellow (XF 4 Tamiya); which is similar to the 'yellow zinc chromate' of the original.

23. The original instrument panel is levelled off. We cement the first photo etch part that will take the instruments.

24. A second part with cut-out dials, is cemented to it with white cement.

25. The photo etch parts are removed with a special cutter.

26. The instrument dials glass is made with Kristal Klear, a sort of gloss varnish.

27. The first stage of the weathering consists of making scratches using a brillo pad or a brush and dark green and aluminium paint.

28. The details are highlighted with Prince August paints.

29. A burnt umber oil is used to reinforce the contrasts.

30. Dust is simulated with the help of different makes of pigments which are then fixed after having been applied with white spirit.

31, 32 and 33. The final result. The photo etch seat and pre-painted straps add much to this cockpit.

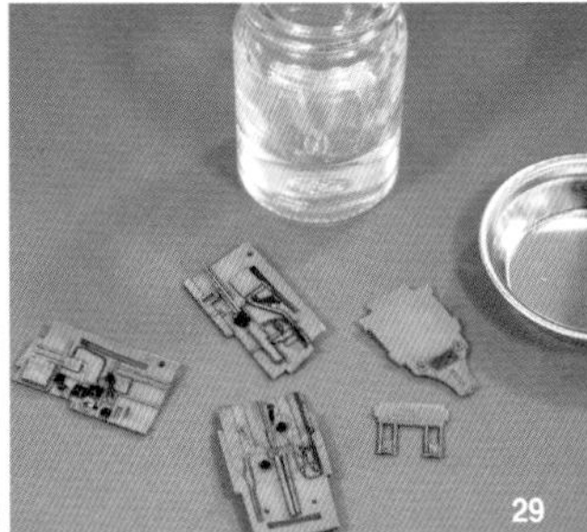

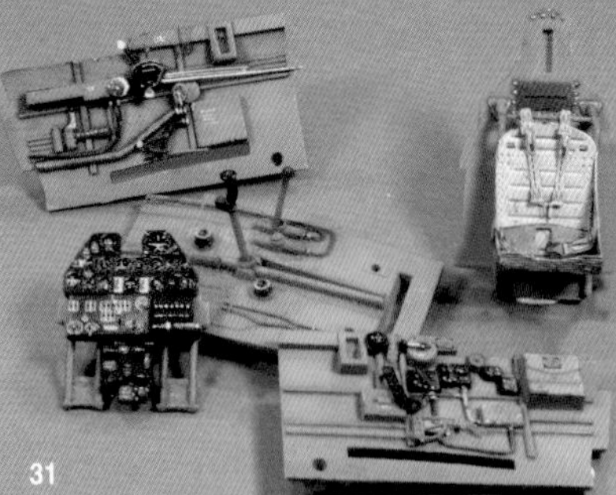

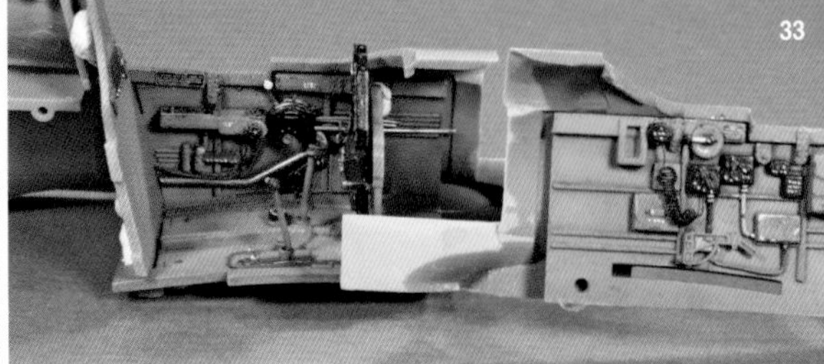

3. Assembly

As with most kits made by this brand, the joins are rather good, except for the wing fillet.

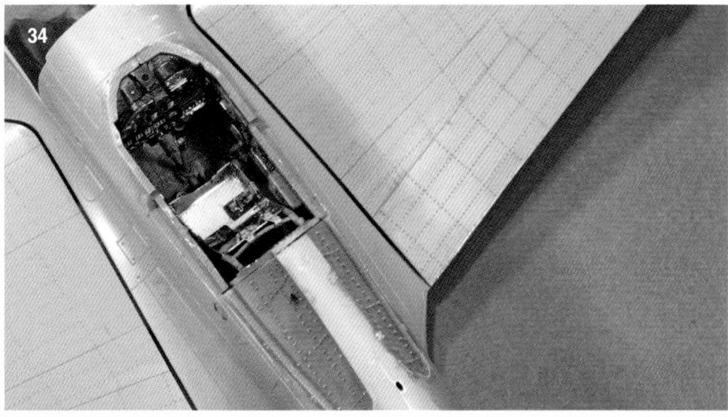

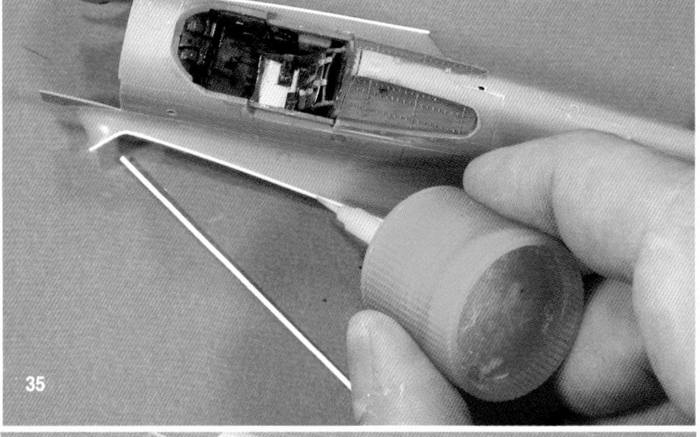

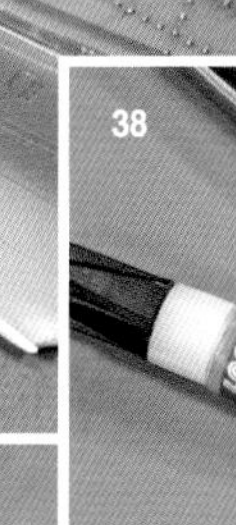

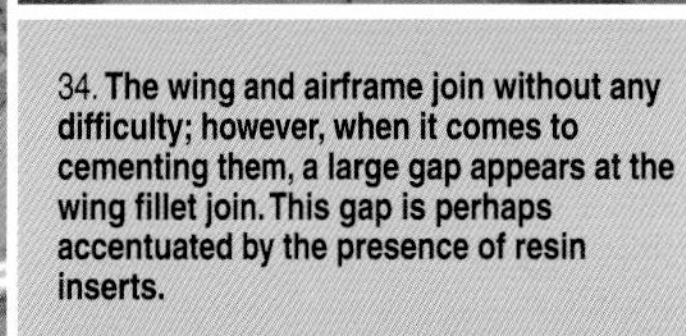

34. The wing and airframe join without any difficulty; however, when it comes to cementing them, a large gap appears at the wing fillet join. This gap is perhaps accentuated by the presence of resin inserts.

35 and 36. To remove this gap, we place a strip of plastic sheet of the correct width along the wing fillet and airframe using liquid cement.

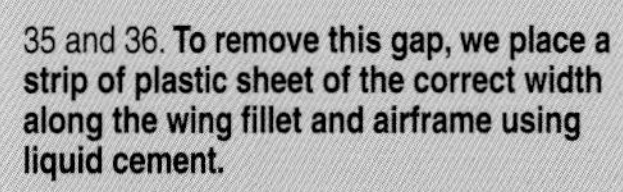

37. This allows us to fill in the gap without difficulty.

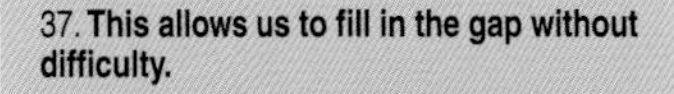

38. The undercarriage legs are detailed using the Eduard sheet. The brake hose is made from tin wire and cemented with cyanoacrylate. This material is easily bendable and allows us to make curves often more realistic than copper wire.

39. The side windows stick out of their emplacement where in reality they were flush with the fuselage. They are, therefore, sanded down with fairly fine sandpaper and water (1,000).

40. **The remaining joins are filled with putty and sanded down with the same materials.**

41. **To regain their initial transparency, we need to sand them down with progressive grains of sandpaper (from 1,000 to 2,000). The windows are then polished with pieces of 3,600 to 12,000 grain Micromesh emery paper. Lastly, we use Tamiya Compound. (a compound that can be polished).**

42 and 43. **The windows have become transparent again and fit the fuselage perfectly.**

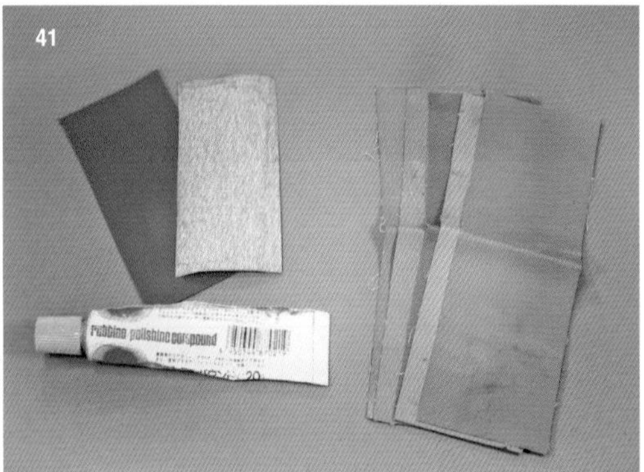

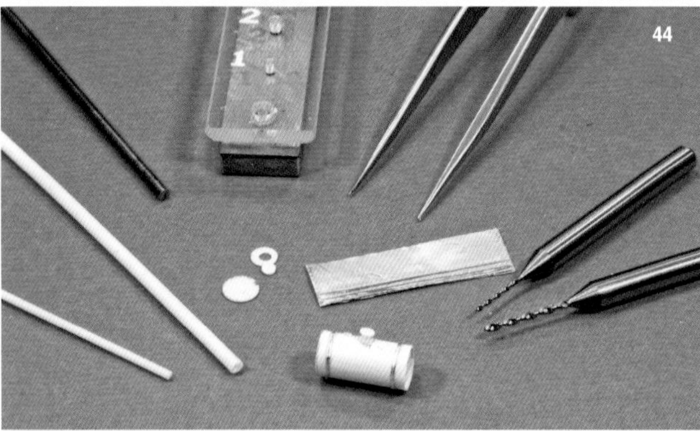

44. **Before painting, we prepare the different parts of the fire wall we are going to make, notably the oil tank.**

45. **The fire wall is a piece of shaped plastic sheet. The emplacements of the various reinforcing and plates are drawn on with a felt tip. The added parts are provisionally placed to ascertain**

46 and 47. **The reinforcing and plates are added. Holes are drilled beforehand for the engine mounting arms and cables**

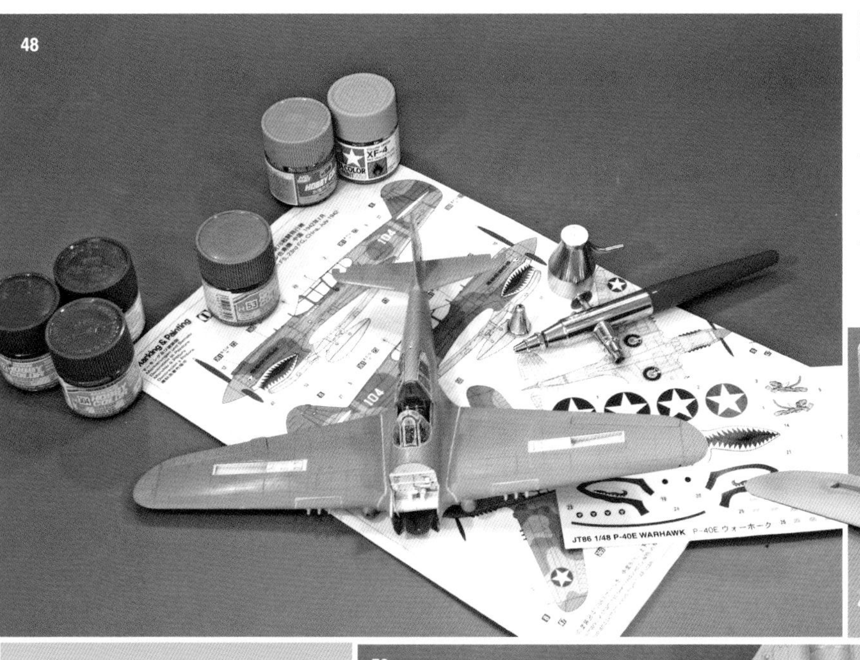

4 Painting and weathering

The olive drab is the base colour of most American aircraft. It is a surprising shade to work with. Actually, depending on the weathering colours used, it is possible to obtain very different camouflage shades, as you can see with the P-40N in the following chapter.

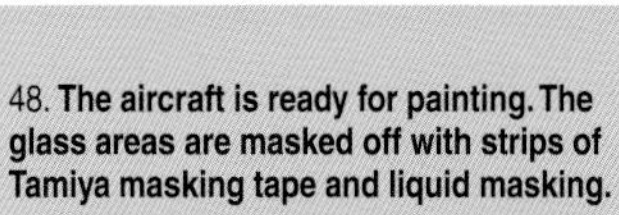

48. The aircraft is ready for painting. The glass areas are masked off with strips of Tamiya masking tape and liquid masking.

49. The gun bays and engine compartment are painted in the right shades then masked.

50. The undercarriage wells are masked with wet kitchen paper. The machine gun compartment is masked with Tamiya masking tape cut to the correct size.

51. Aluminium base paint is applied on the walking areas. Small amounts of liquid masking are applied before painting and will give a nice weathering effect.
Lower surfa

52. The lower surfaces of the aircraft are completely in Neutral Gray/Gunze H 53.

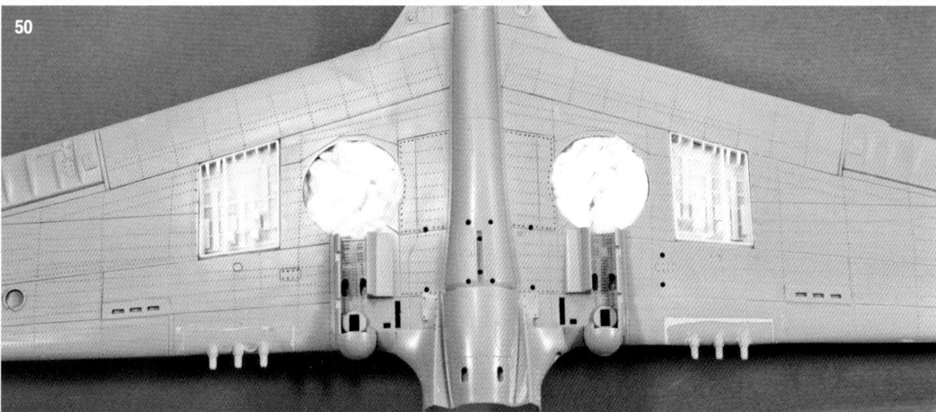

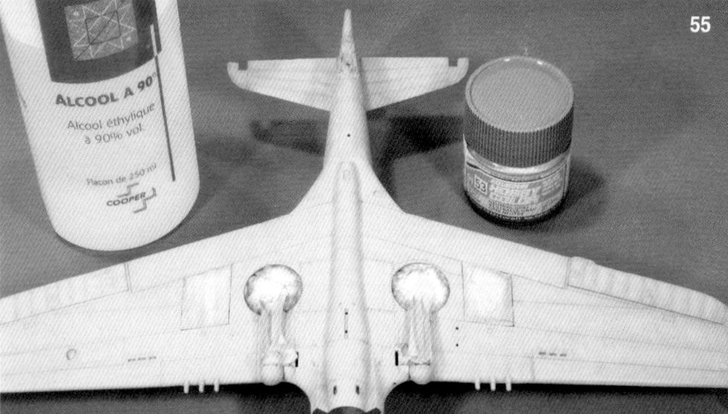

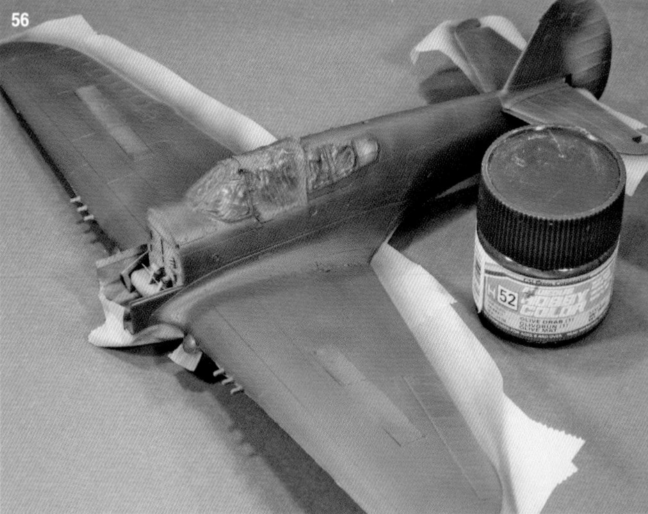

53. The base shade is firstly lightened with white. One should try and vary the effects (inside of panels, structural lines, streaks in the direction of the relative airflow etc.)

54. We use the same method with a dark mix, making sure we make it look as natural as possible.

55. We shade off with an extremely diluted (95%) pure shade.

56. After masking the lower surfaces, the upper wing surfaces are painted in olive drab (Gunze H 65)

57. The painting is done in the same way as the lower wing surfaces. The olive drab is re-worked with yellow (Gunze H 34) and green black (Gunze H 65).

58. Repainted areas (frequent in theatres of operation) are simulated with a darker olive drab (Gunze H 65).

59. After a coat of gloss varnish, the decals are applied using softening products from the Microscale range. The decals come from the Hasegawa box.

60. The propeller decals are cut out with a punch.

61. A second coat of gloss varnish seals the decals. Matt X-Tracrylix varnish gives the model a more operational aspect.

62. Initial weathering is carried out with an airbrush on the lower wing surfaces following the same techniques as those shown in the P-40B article.

63. Exhaust marks are made with a cream colour (Gunze H 318) highlighted in brown. The vertical streaks are made using a fine brush with acrylic sepia ink and matt varnish, both of which are very diluted.

64. The undercarriage is added at this stage. The airframe is finished; the engine detailing can therefore begin.

65. The Eduard photo etched undercarriage door parts add a lot of realism.

5. The engine

P-40 super detailing kits cannot be commercially bought. Only the engine is supplied by Aires (or CMK). All the rest is for you to do (fire wall and all associated elements, engine mounting arms, filters, engine hood plates etc.)

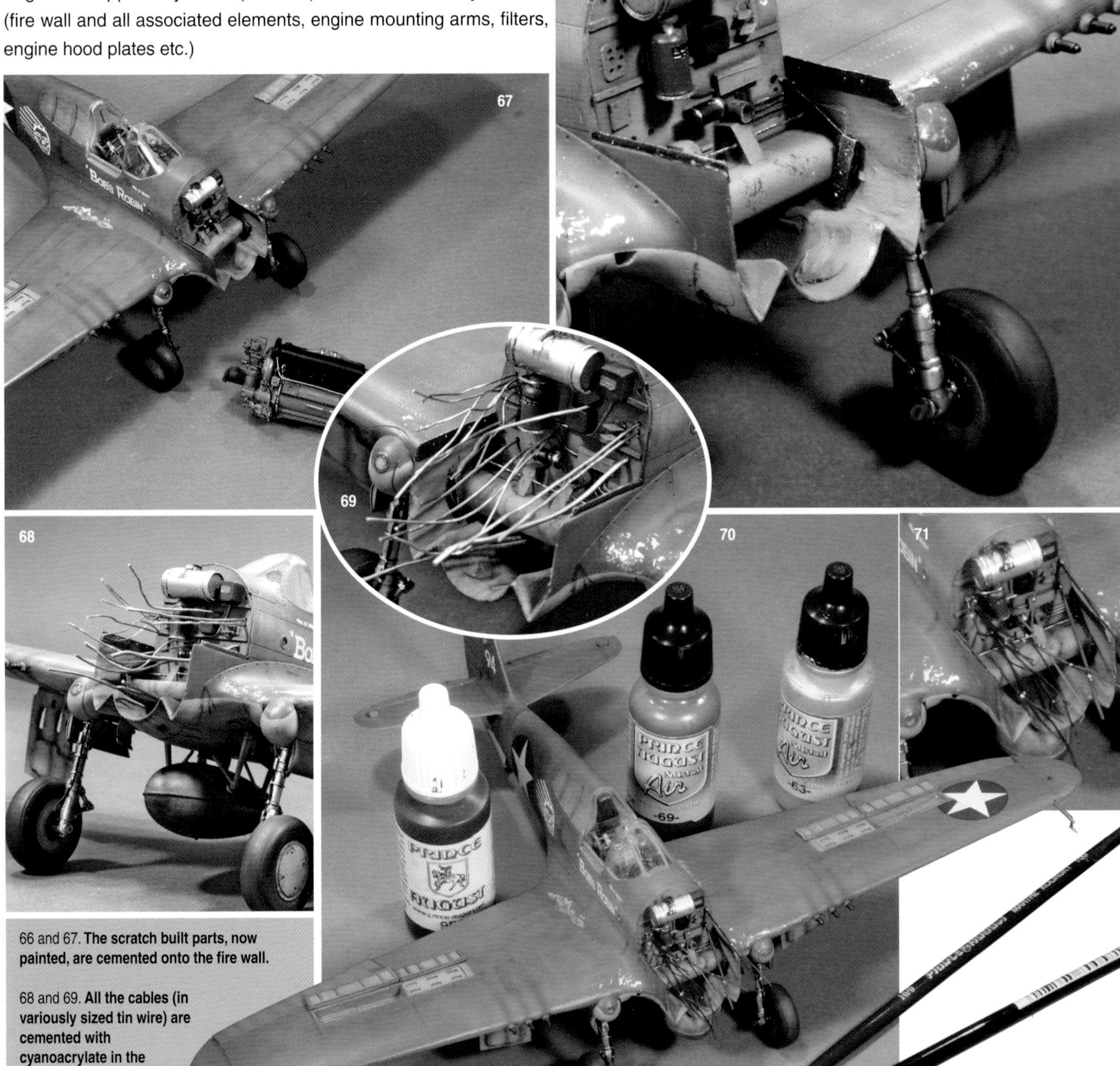

66 and 67. **The scratch built parts, now painted, are cemented onto the fire wall.**

68 and 69. **All the cables (in variously sized tin wire) are cemented with cyanoacrylate in the previously drilled out holes.**

70 and 71. **All the cables are painted black or in tire grey. All the small touches of copper or aluminium simulate the different connections, in a trompe l'oeil.**

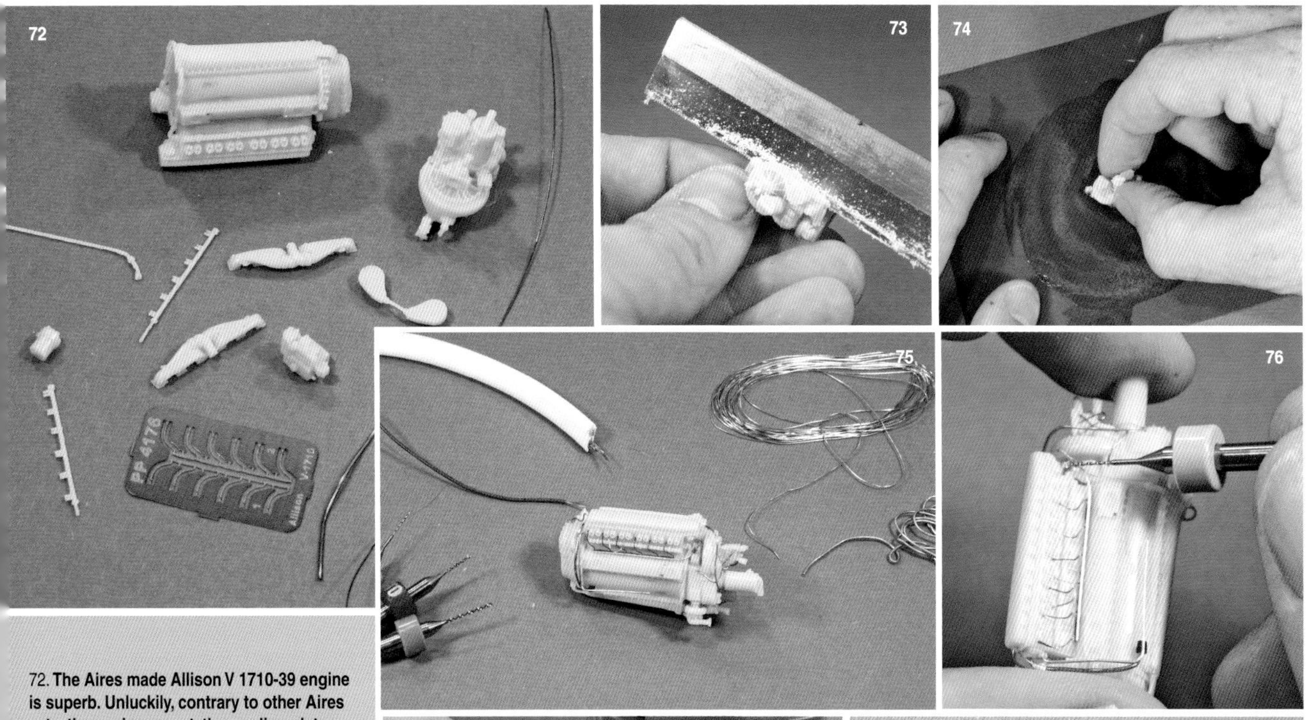

72. The Aires made Allison V 1710-39 engine is superb. Unluckily, contrary to other Aires sets, the engine mount, the cowling plates, filters and other elements are not included.

73 et 74. The large parts are cut with a saw. Any excess is removed by sanding. The dust is noxious and it is advisable to wear a mask with a filter during this process. Sanding with wet sand paper reduces the amount of dust.

75. The engine is detailed with copper or tin wire depending on the effect sought after.

76. To install a hose, it is better to drill a small hole of the same diameter as the chosen wire.

77. We put a little cyanoacrylate cement in the hole and we can install the metal wire. The cementing will be solid and precise.

78. The Allison engine body was engine grey, here neutral grey with a little black. The shade is applied with the diluting product of the same brand to conserve a satin aspect.

79. The details are touched up with Prince August shades, and notably with the Prince August Air metallic shades that hold to the brush very well.

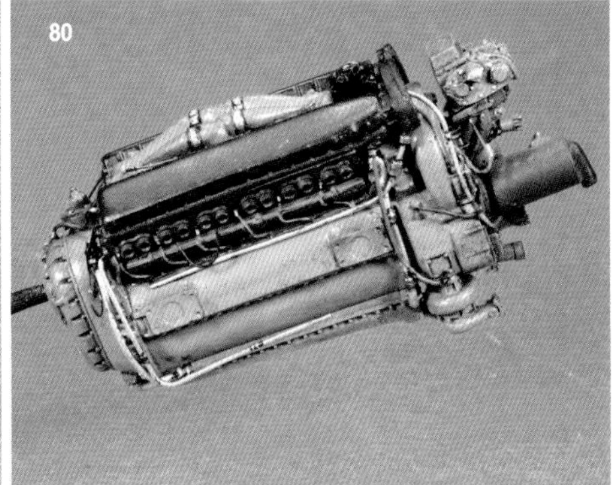

80. A sepia wash, diluted with white spirit is applied to the engine.

81.The engine mount has been simplified in order to support the weight of the resin engine. Metallic rods are used to avoid torsion.

82. The engine mount arms are painted in interior green.

83. The kit's exhausts are cut out. Their receptacle on the engine is thermo shaped on a template made from thick plastic sheet.

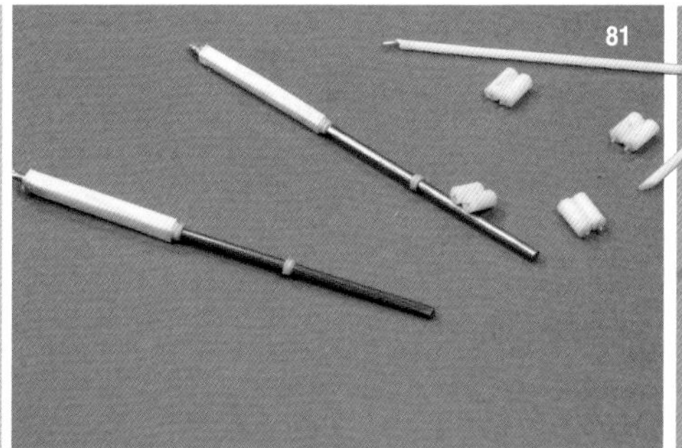

81

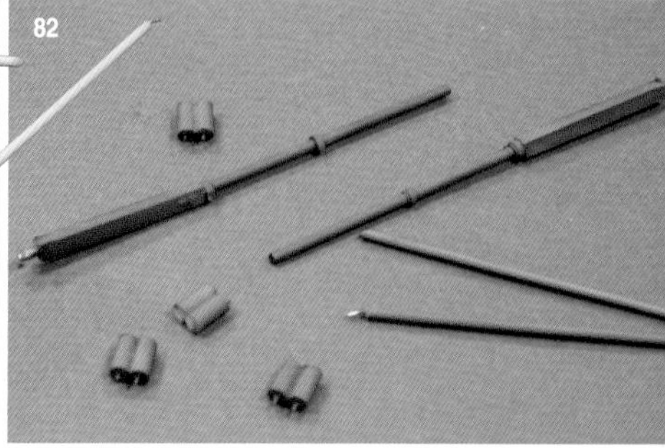

82

83

84. The filters are made with the help of tubes and the part from the Hasegawa kit. The grills come from the Eduard sheet.

85. The assembled filters. A piece of aluminium sheet forms the rear part.

86. Painting these parts is done separately. The exhausts are painted brown and rubbed with rust colour pigments of different brands. It is fiddly to place the engine on its arms and to link everything up to the firewall.

87. The filters are linked to the engine mount by small pieces of plastic sheet. The fixing plates are simulated using photo etch off-cuts.

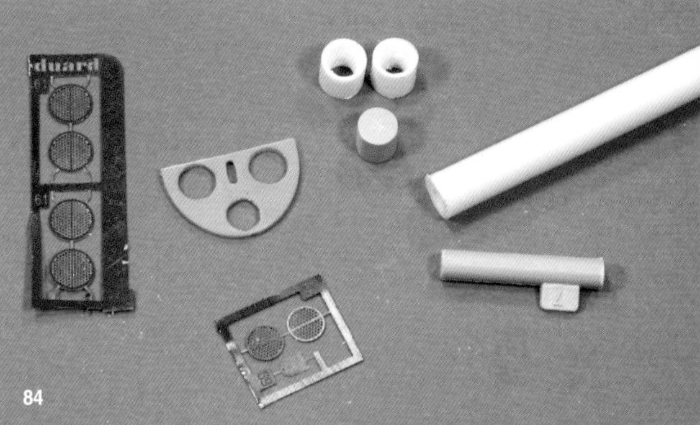

84

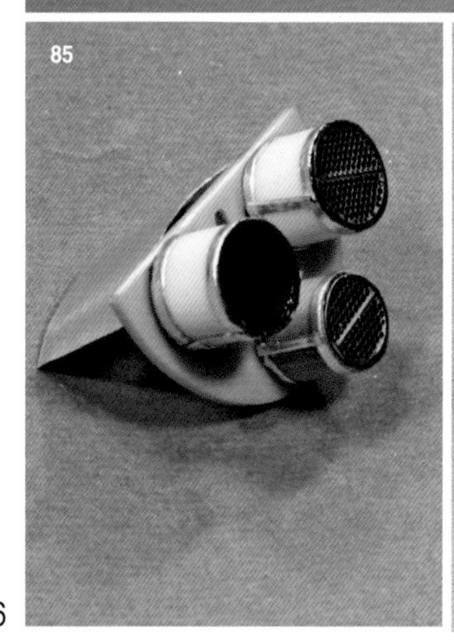

85

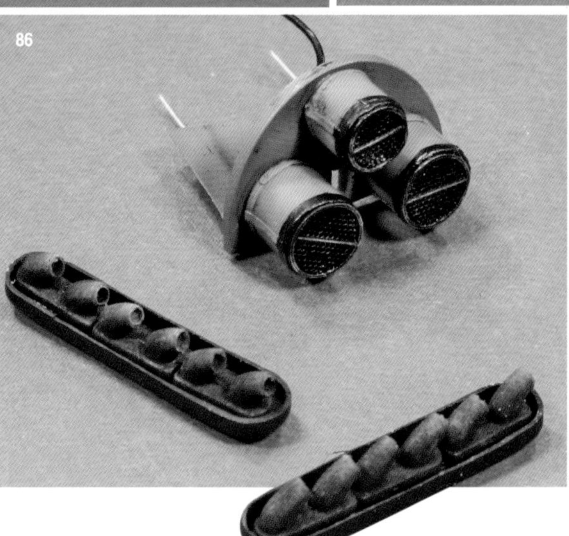

86

87

"Bob's Rob

"BOB'S ROBIN"
94

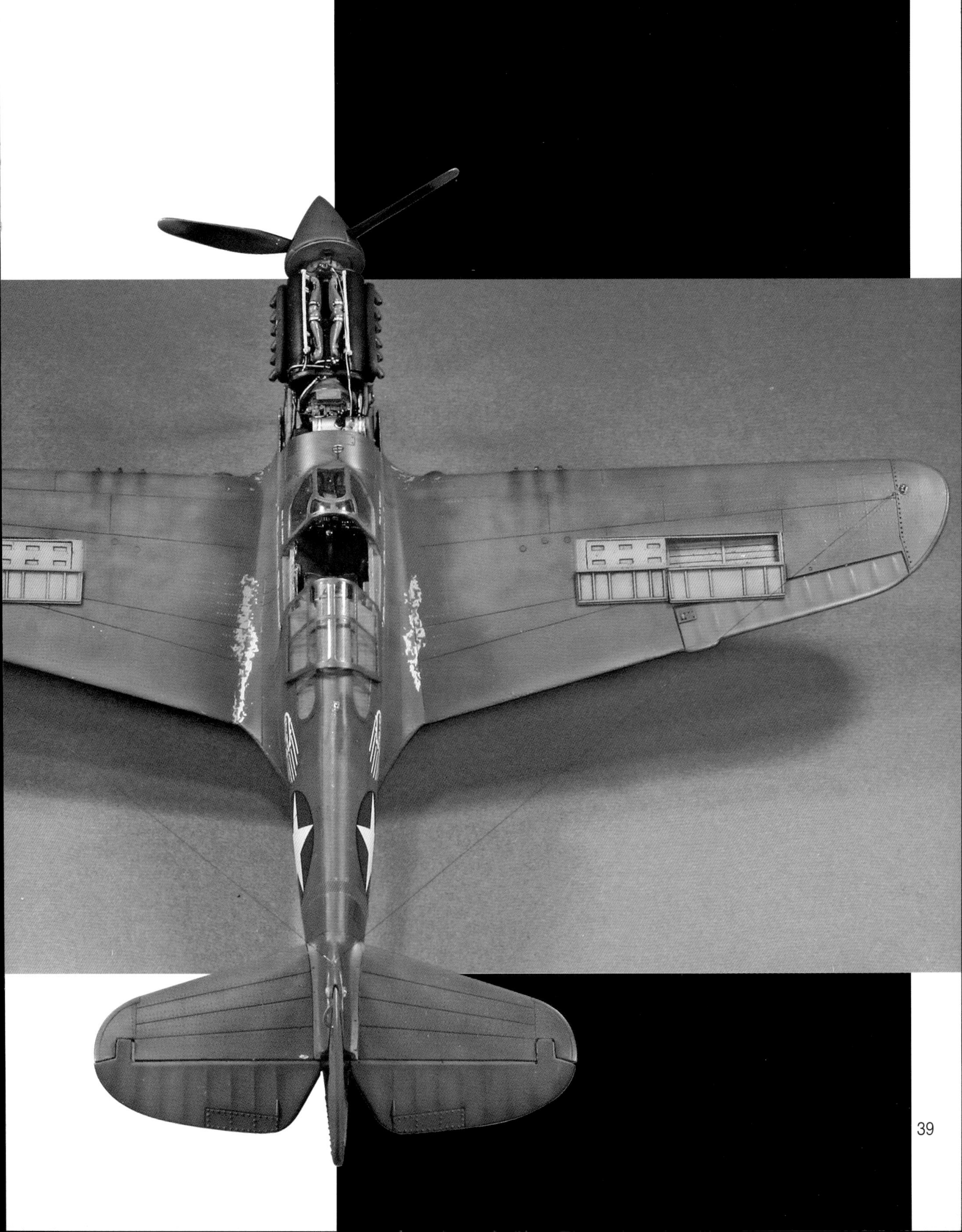

94
'BOB'S ROBIN'

94
"BOB'S ROBIN"

CURTISS P-40B

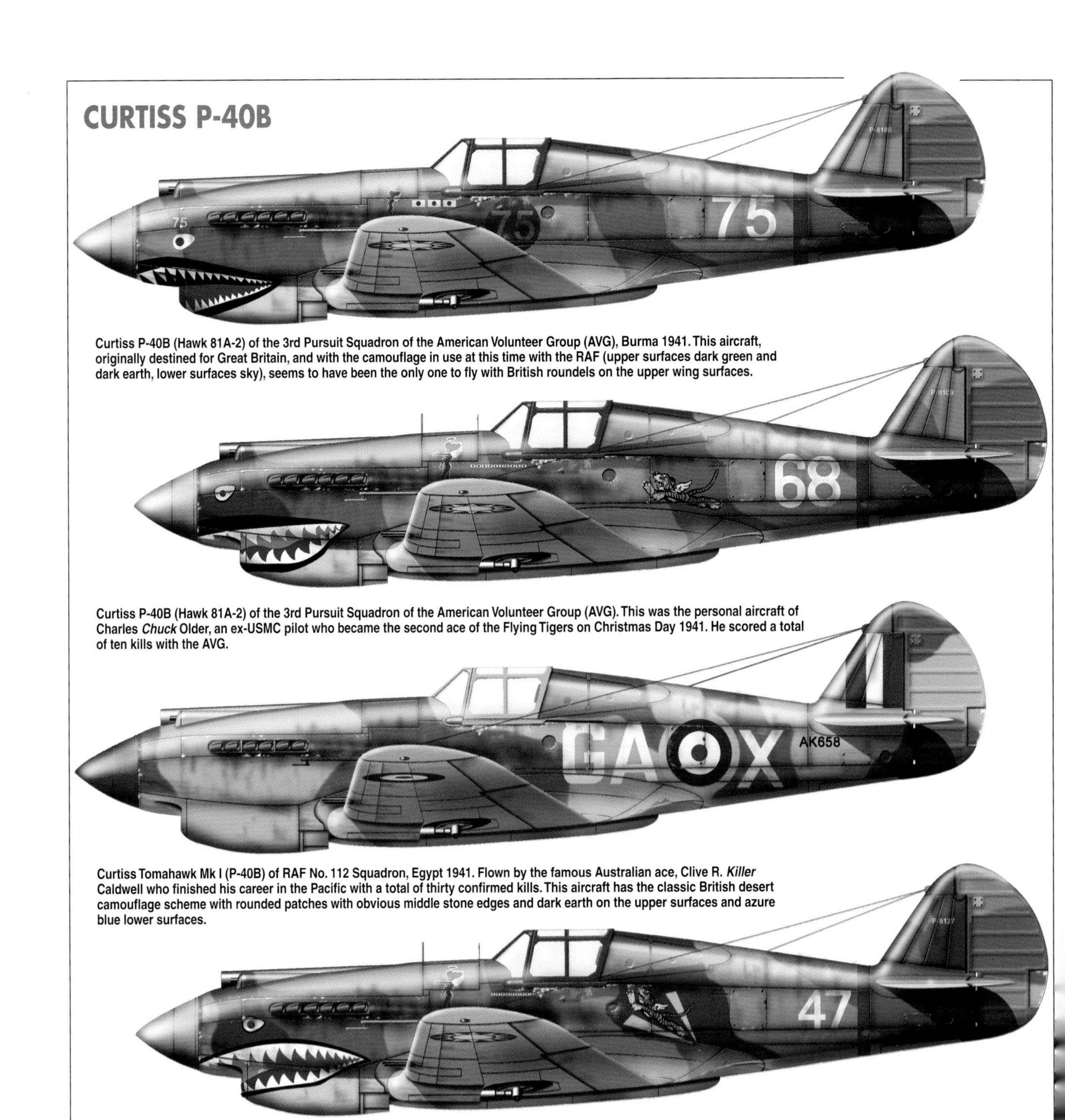

Curtiss P-40B (Hawk 81A-2) of the 3rd Pursuit Squadron of the American Volunteer Group (AVG), Burma 1941. This aircraft, originally destined for Great Britain, and with the camouflage in use at this time with the RAF (upper surfaces dark green and dark earth, lower surfaces sky), seems to have been the only one to fly with British roundels on the upper wing surfaces.

Curtiss P-40B (Hawk 81A-2) of the 3rd Pursuit Squadron of the American Volunteer Group (AVG). This was the personal aircraft of Charles *Chuck* Older, an ex-USMC pilot who became the second ace of the Flying Tigers on Christmas Day 1941. He scored a total of ten kills with the AVG.

Curtiss Tomahawk Mk I (P-40B) of RAF No. 112 Squadron, Egypt 1941. Flown by the famous Australian ace, Clive R. *Killer* Caldwell who finished his career in the Pacific with a total of thirty confirmed kills. This aircraft has the classic British desert camouflage scheme with rounded patches with obvious middle stone edges and dark earth on the upper surfaces and azure blue lower surfaces.

Curtiss P-40B (Hawk 81-A2) of the 3rd Pursuit Squadron of the American Volunteer Group (AVG), China, 1942. This plane was flown by Robert Tharp Smith who was placed in command of the 3rd PS (Hell's Angels) in 1942. Smith scored 8.75 kills with the Flying Tigers and was decorated twice for this by the Chinese government. The AVG was disbanded on 4 July 1942 and integrated into the USAAF.

CURTISS P-40E

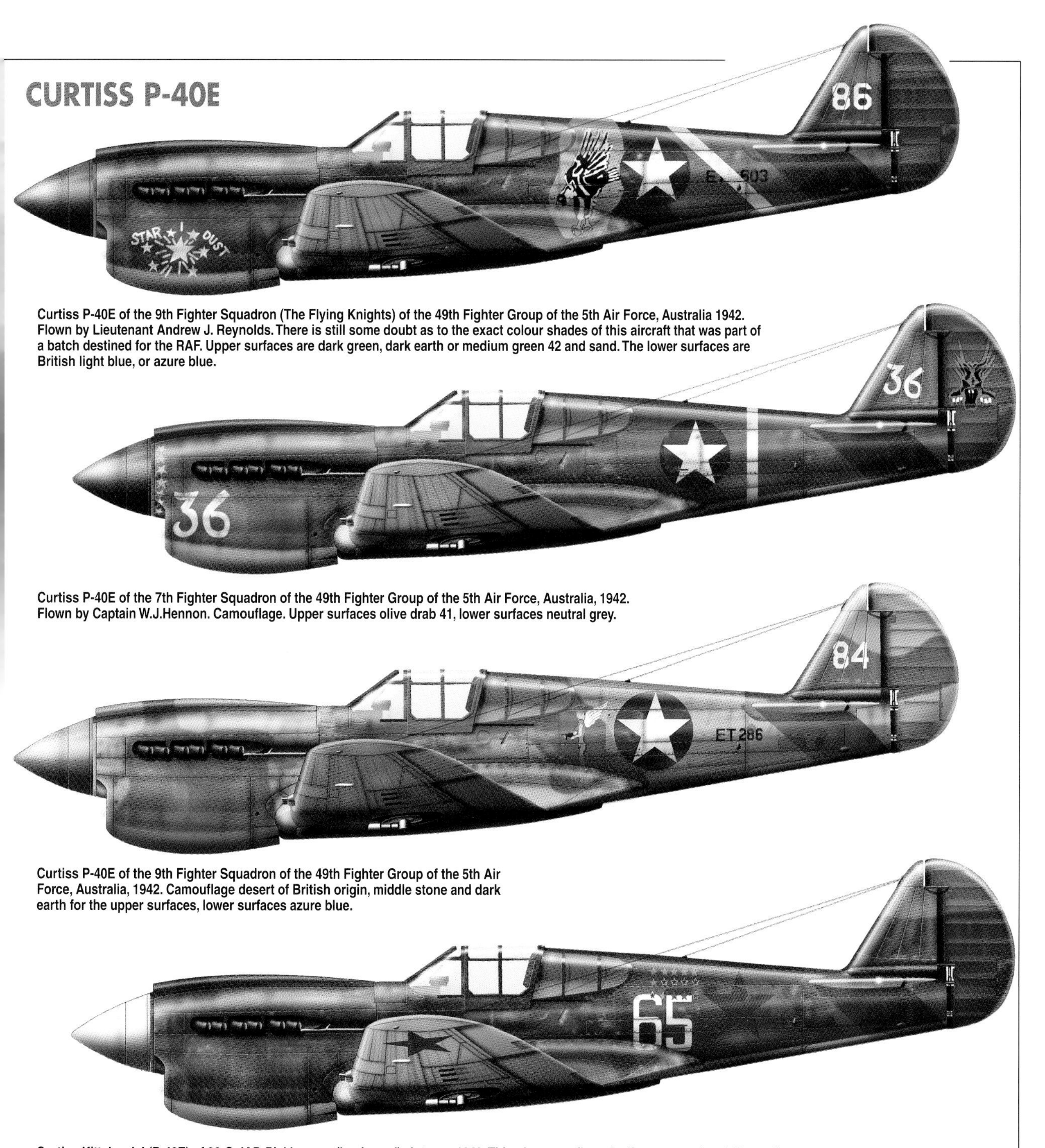

Curtiss P-40E of the 9th Fighter Squadron (The Flying Knights) of the 49th Fighter Group of the 5th Air Force, Australia 1942. Flown by Lieutenant Andrew J. Reynolds. There is still some doubt as to the exact colour shades of this aircraft that was part of a batch destined for the RAF. Upper surfaces are dark green, dark earth or medium green 42 and sand. The lower surfaces are British light blue, or azure blue.

Curtiss P-40E of the 7th Fighter Squadron of the 49th Fighter Group of the 5th Air Force, Australia, 1942. Flown by Captain W.J.Hennon. Camouflage. Upper surfaces olive drab 41, lower surfaces neutral grey.

Curtiss P-40E of the 9th Fighter Squadron of the 49th Fighter Group of the 5th Air Force, Australia, 1942. Camouflage desert of British origin, middle stone and dark earth for the upper surfaces, lower surfaces azure blue.

Curtiss Kittyhawk I (P-40E) of 29.GvIAP. Piekhanovo (Leningrad), Autumn 1942. This plane was flown by lieutenant colonel Alexander Matveyev. He scored six individual kills (yellow stars above the number) and eight associated kills (red stars). He finished the war as commander of this Guards fighter regiment. Supplied by the British, this plane has kept its original British camouflage (upper surfaces brown and dark green, lower surfaces sky).

CURTISS P-40F

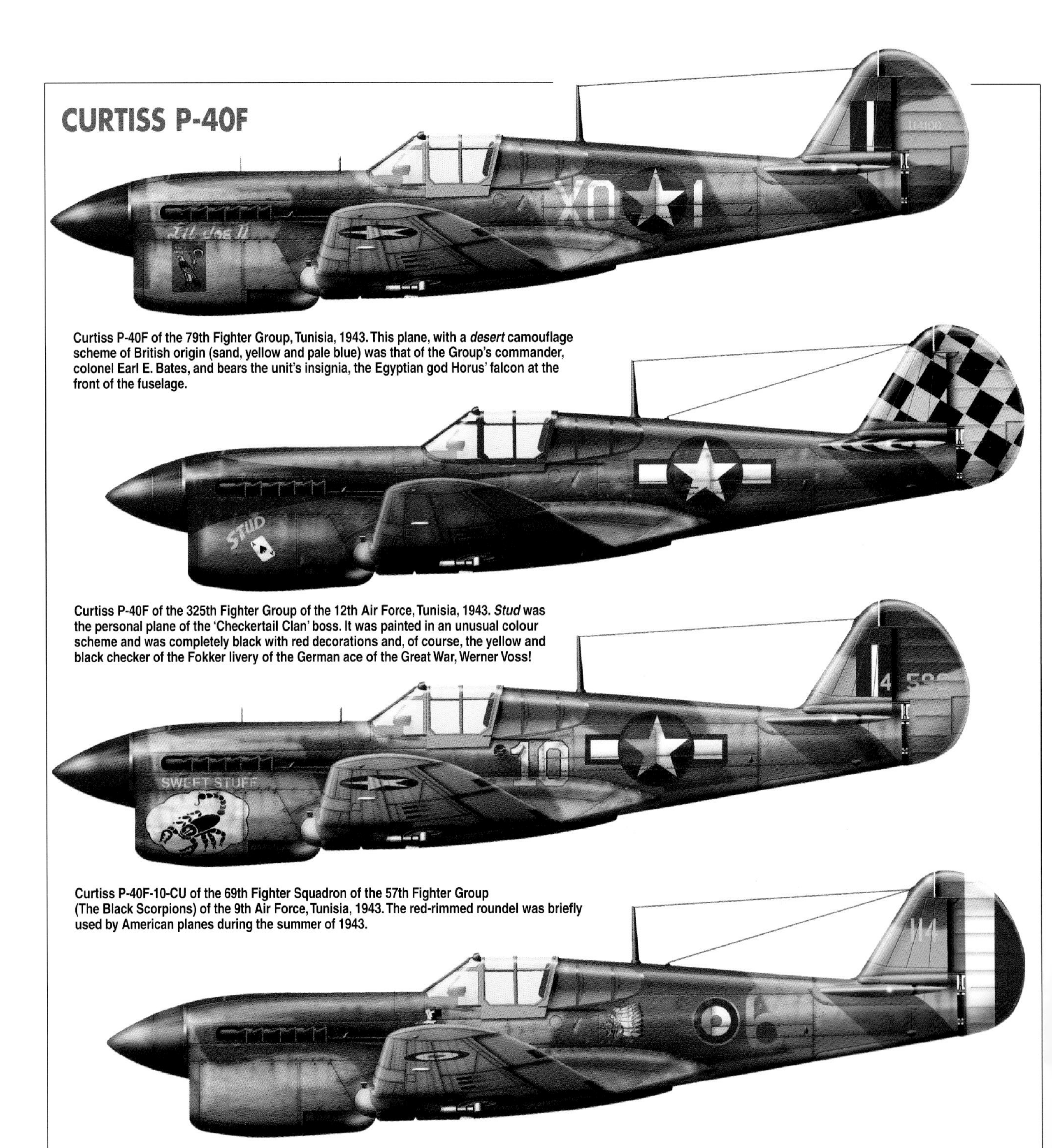

Curtiss P-40F of the 79th Fighter Group, Tunisia, 1943. This plane, with a *desert* camouflage scheme of British origin (sand, yellow and pale blue) was that of the Group's commander, colonel Earl E. Bates, and bears the unit's insignia, the Egyptian god Horus' falcon at the front of the fuselage.

Curtiss P-40F of the 325th Fighter Group of the 12th Air Force, Tunisia, 1943. *Stud* was the personal plane of the 'Checkertail Clan' boss. It was painted in an unusual colour scheme and was completely black with red decorations and, of course, the yellow and black checker of the Fokker livery of the German ace of the Great War, Werner Voss!

Curtiss P-40F-10-CU of the 69th Fighter Squadron of the 57th Fighter Group (The Black Scorpions) of the 9th Air Force, Tunisia, 1943. The red-rimmed roundel was briefly used by American planes during the summer of 1943.

Curtiss P-40F of the Groupe de Chasse 2/5 *La Fayette*, Algeria, 1943. Flown by sergeant Jean Gisclon. In memory of the *Escadrille La Fayette* Made up of American volunteers that fought alongside the French during the Great War, this Armée de l'Air group, reconstituted in North Africa, was the first to be equipped with American fighters. These aircraft had British camouflage schemes and bore the markings of their new owners, notably the two unit insignia, the flying stork of the SPA 167 squadron and the Indian head of the N 123.

CURTISS P-40N

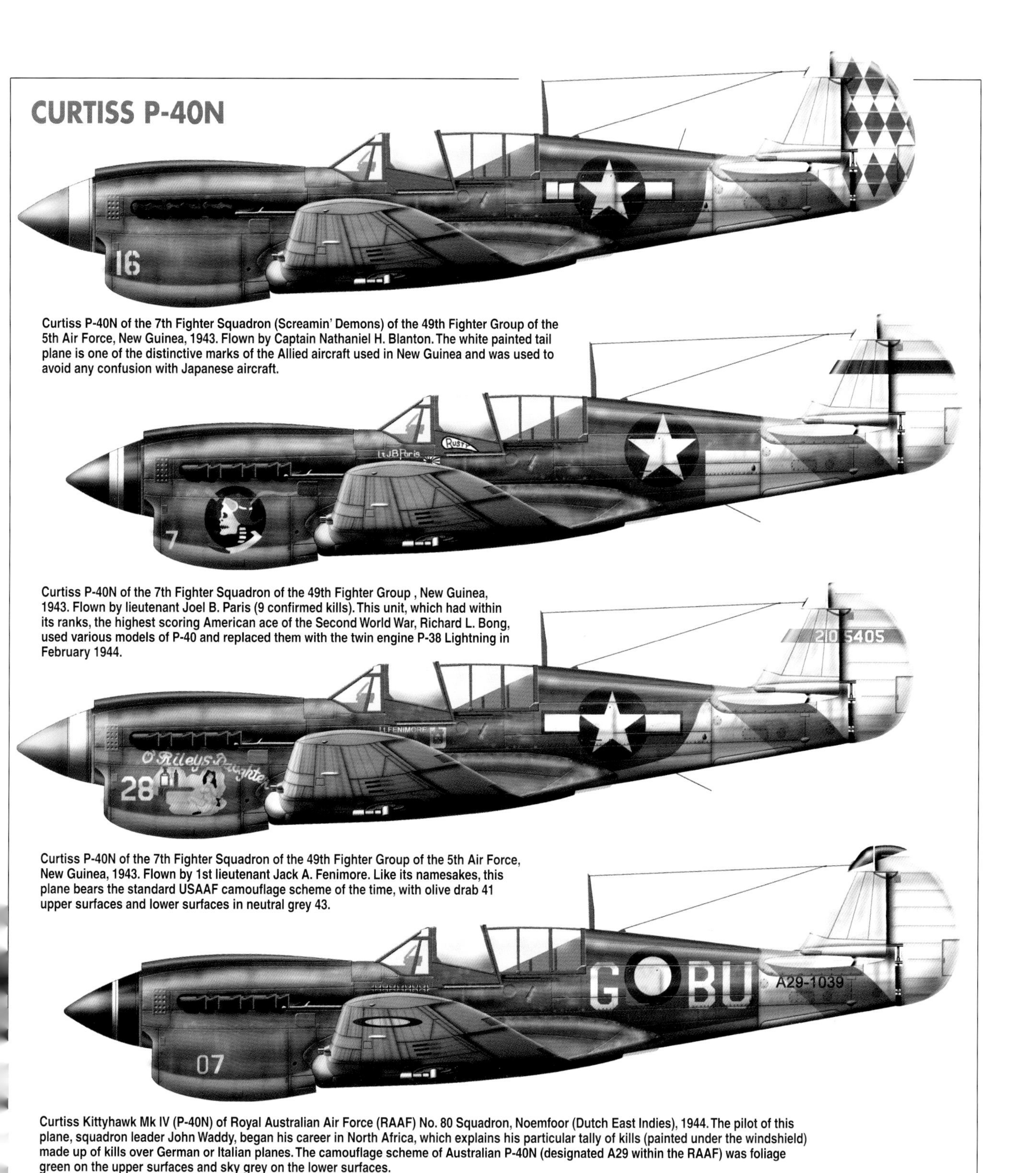

Curtiss P-40N of the 7th Fighter Squadron (Screamin' Demons) of the 49th Fighter Group of the 5th Air Force, New Guinea, 1943. Flown by Captain Nathaniel H. Blanton. The white painted tail plane is one of the distinctive marks of the Allied aircraft used in New Guinea and was used to avoid any confusion with Japanese aircraft.

Curtiss P-40N of the 7th Fighter Squadron of the 49th Fighter Group , New Guinea, 1943. Flown by lieutenant Joel B. Paris (9 confirmed kills). This unit, which had within its ranks, the highest scoring American ace of the Second World War, Richard L. Bong, used various models of P-40 and replaced them with the twin engine P-38 Lightning in February 1944.

Curtiss P-40N of the 7th Fighter Squadron of the 49th Fighter Group of the 5th Air Force, New Guinea, 1943. Flown by 1st lieutenant Jack A. Fenimore. Like its namesakes, this plane bears the standard USAAF camouflage scheme of the time, with olive drab 41 upper surfaces and lower surfaces in neutral grey 43.

Curtiss Kittyhawk Mk IV (P-40N) of Royal Australian Air Force (RAAF) No. 80 Squadron, Noemfoor (Dutch East Indies), 1944. The pilot of this plane, squadron leader John Waddy, began his career in North Africa, which explains his particular tally of kills (painted under the windshield) made up of kills over German or Italian planes. The camouflage scheme of Australian P-40N (designated A29 within the RAAF) was foliage green on the upper surfaces and sky grey on the lower surfaces.

CURTISS P-40F

FRENCH ARMEE DE L'AIR *CONVERSION*

1/48 SCALE BUILDING

By Olivier Soulleys

THERE IS NO P-40F AVAILABLE IN 1/48 SCALE, unless Hasegawa keeps going and offers makes us an F or L, why not?
It is, however, for many, the most aesthetic version of the P-40, thanks notably to its smooth cowling. With good references, and good tools, this article shows a simple conversion that does not need any accessories.

The spare parts and elements needed for improvement that will perhaps be used on our P-40.

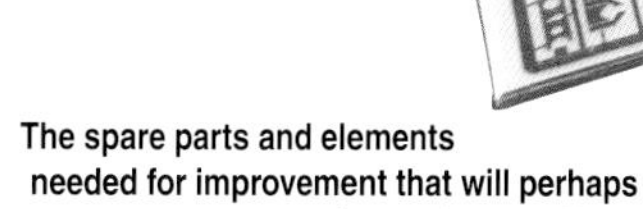

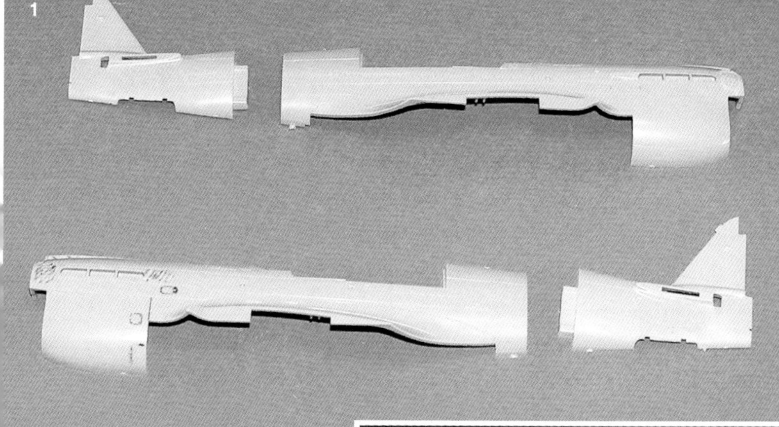

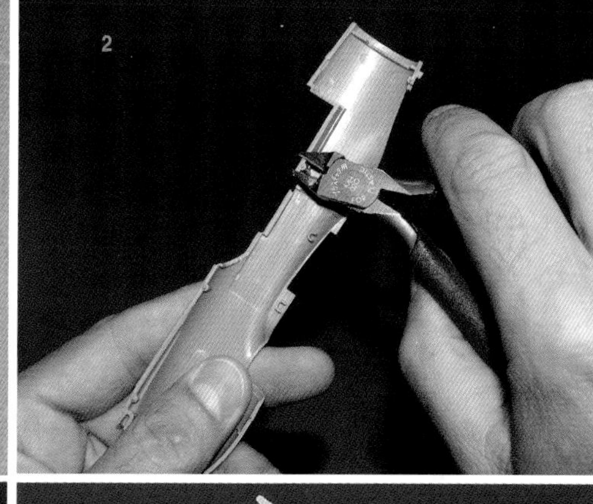

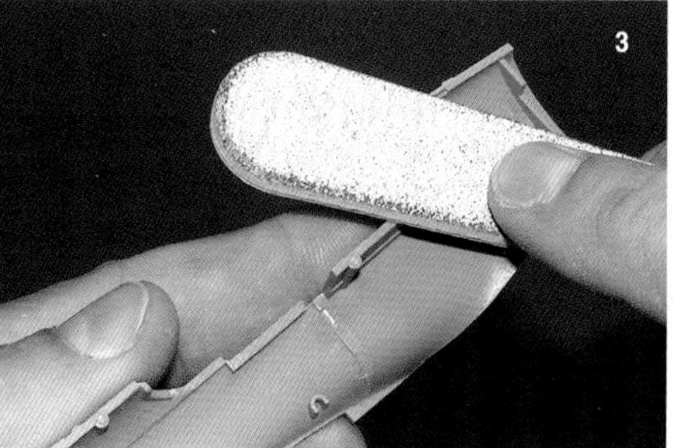

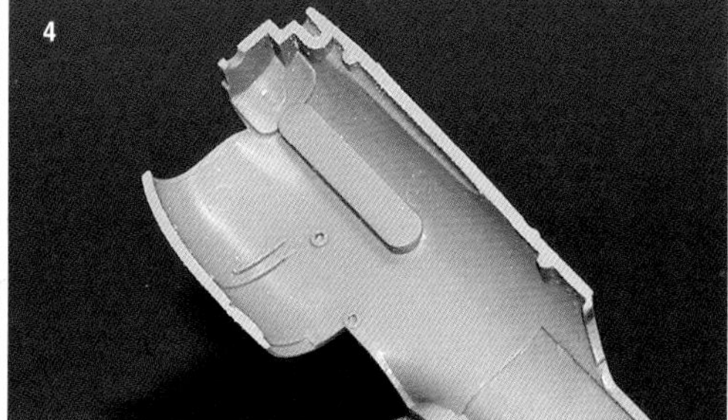

1. The two fuselage halves are ready to be assembled separately.

2. The slugs are removed with a cutter.

3 and 4. The surfaces are sanded down with a rigid file that ensures an even sanding.

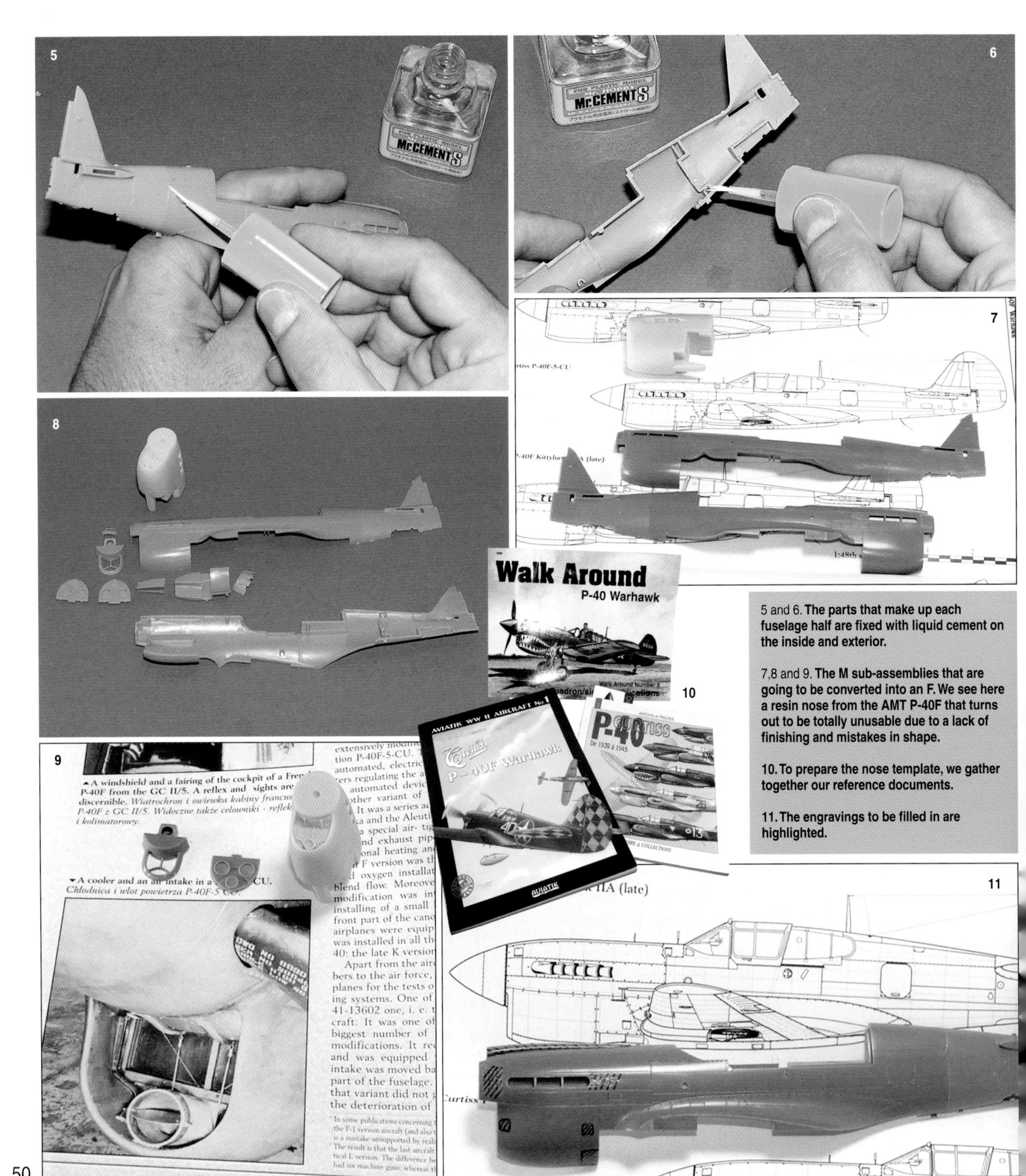

5 and 6. **The parts that make up each fuselage half are fixed with liquid cement on the inside and exterior.**

7,8 and 9. **The M sub-assemblies that are going to be converted into an F. We see here a resin nose from the AMT P-40F that turns out to be totally unusable due to a lack of finishing and mistakes in shape.**

10. **To prepare the nose template, we gather together our reference documents.**

11. **The engravings to be filled in are highlighted.**

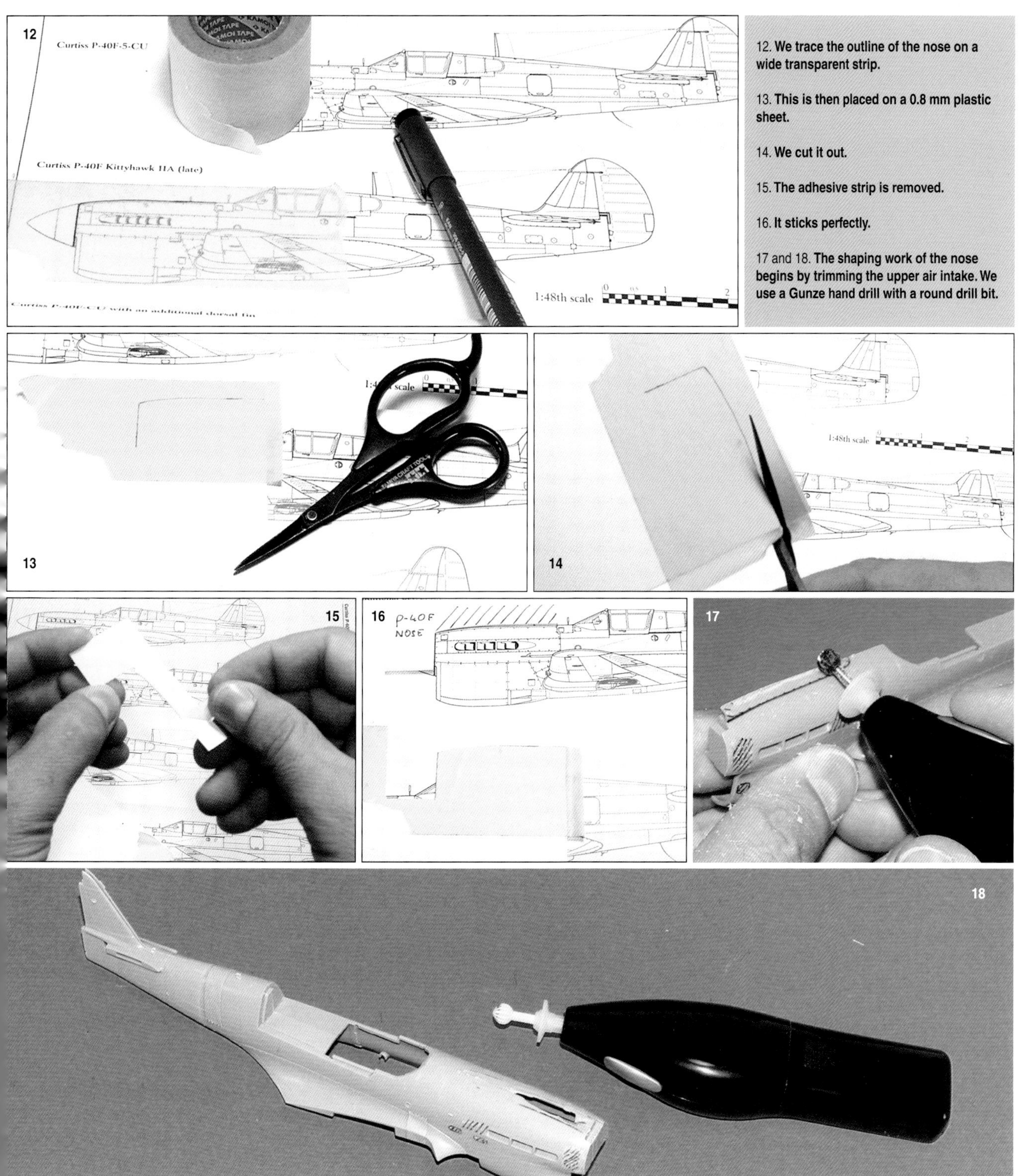

12. **We trace the outline of the nose on a wide transparent strip.**

13. **This is then placed on a 0.8 mm plastic sheet.**

14. **We cut it out.**

15. **The adhesive strip is removed.**

16. **It sticks perfectly.**

17 and 18. **The shaping work of the nose begins by trimming the upper air intake. We use a Gunze hand drill with a round drill bit.**

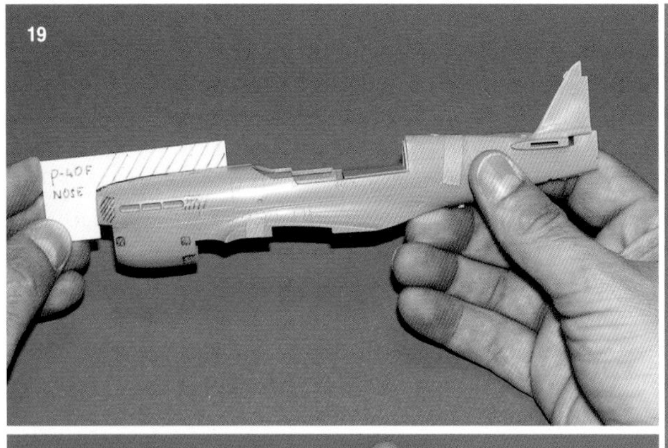

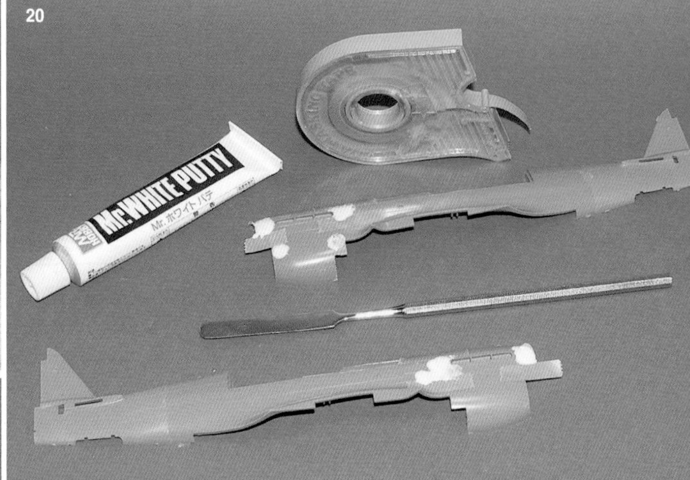

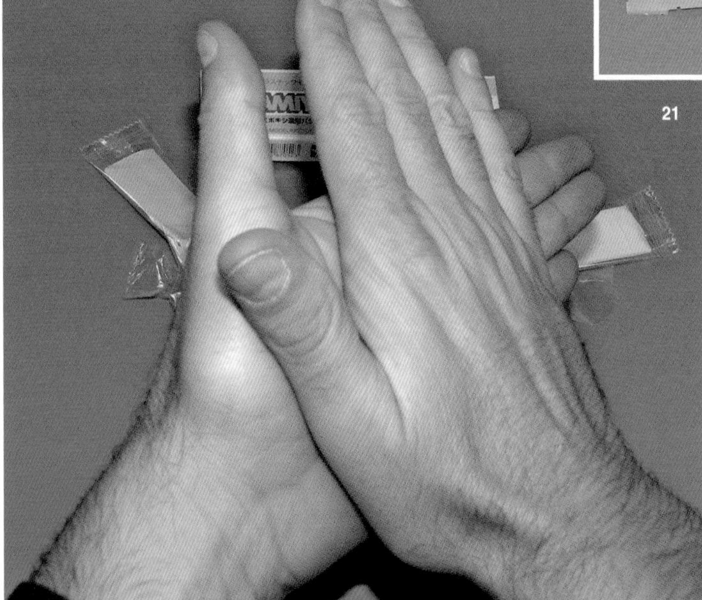

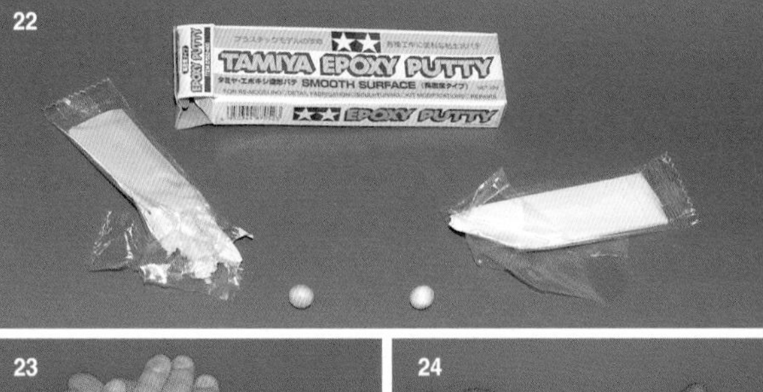

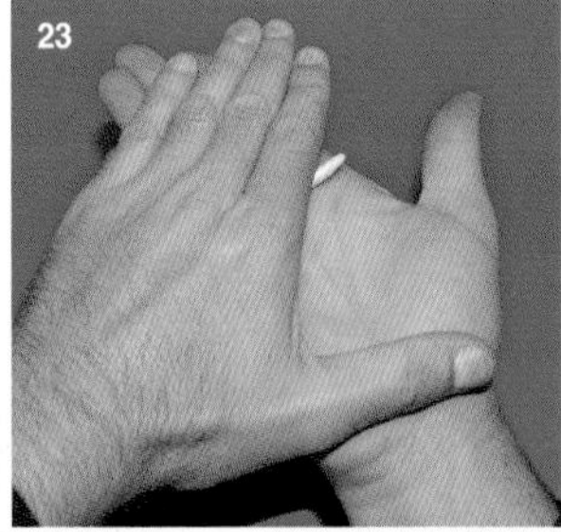

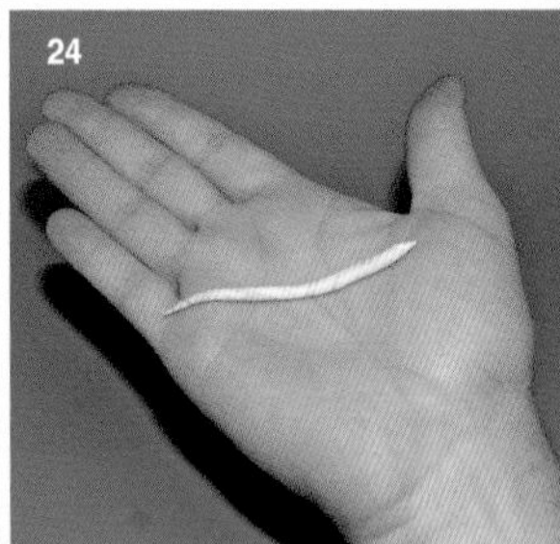

19. **A first check reveals that we need to fill in part of the cowling with putty.**

20. **We make the most of this to also fill in the structures that do not exist on the version we wish to build.**

21 and 24. **Using two equal sized balls of epoxy putty, we prepare the putty that will fill in the cowling.**

25. **After marking the areas concerned, we add the putty.**

26. **As the round drill bit has made a hole in the top of the cowling, we cement a block inside that the double component putty can rest on.**

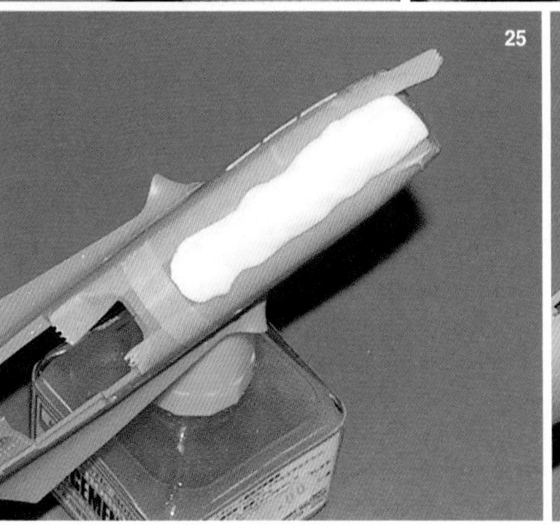

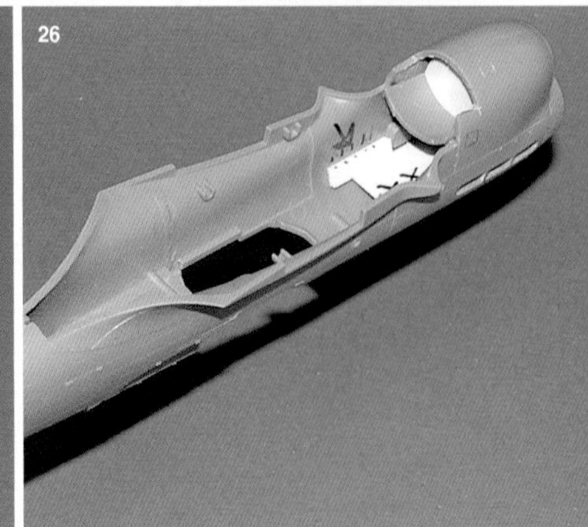

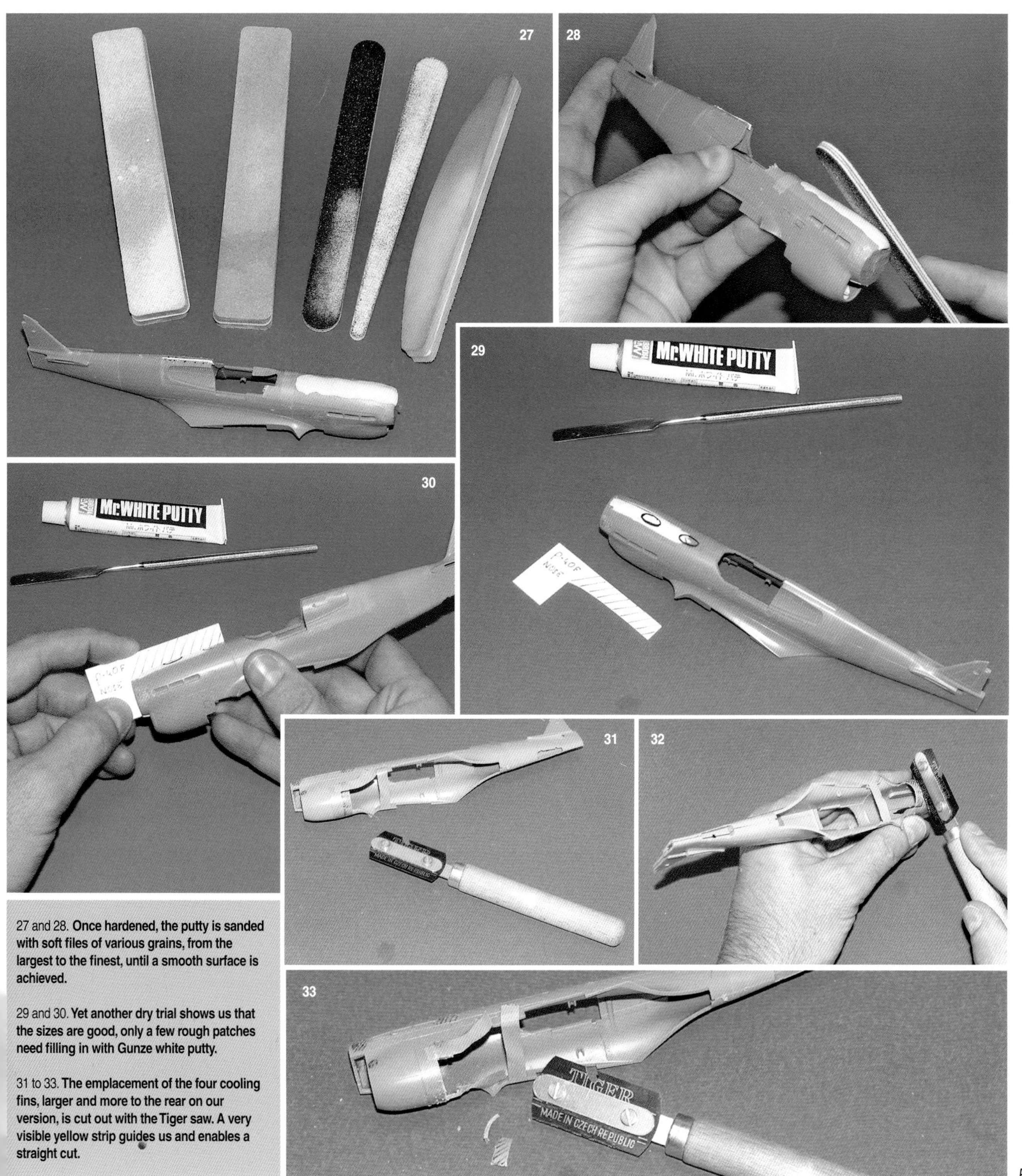

27 and 28. **Once hardened, the putty is sanded with soft files of various grains, from the largest to the finest, until a smooth surface is achieved.**

29 and 30. **Yet another dry trial shows us that the sizes are good, only a few rough patches need filling in with Gunze white putty.**

31 to 33. **The emplacement of the four cooling fins, larger and more to the rear on our version, is cut out with the Tiger saw. A very visible yellow strip guides us and enables a straight cut.**

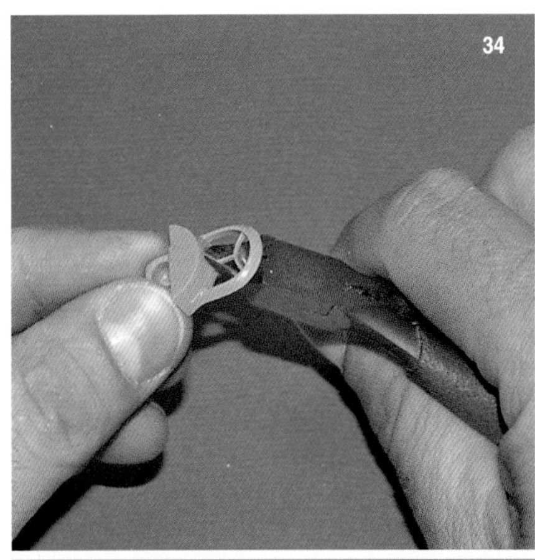

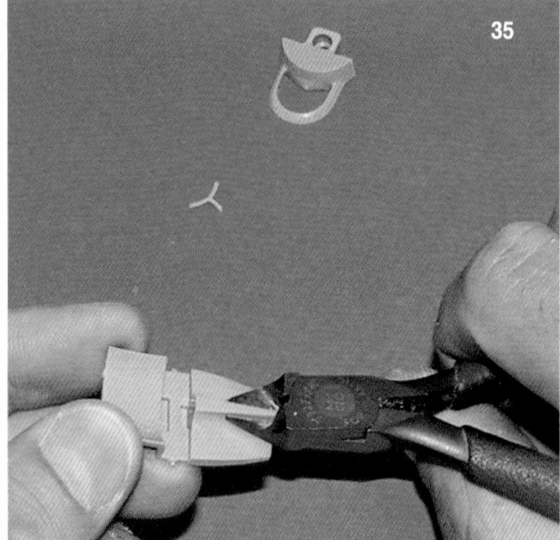

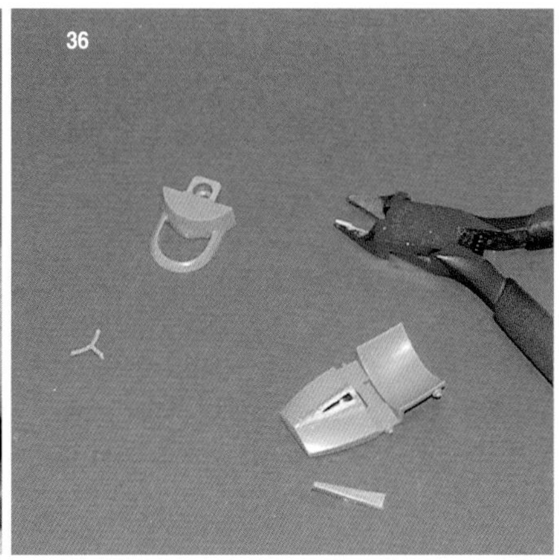

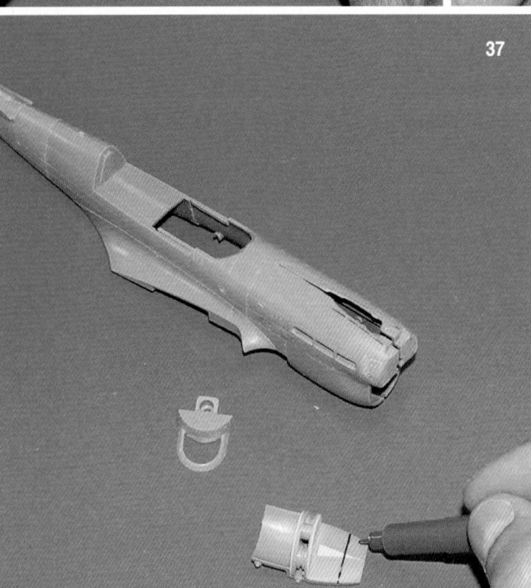

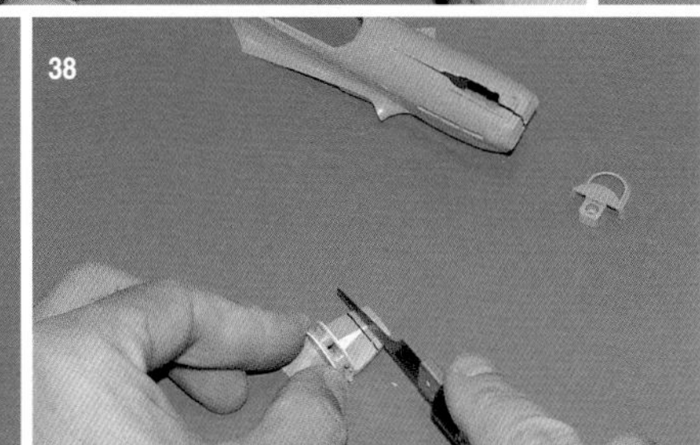

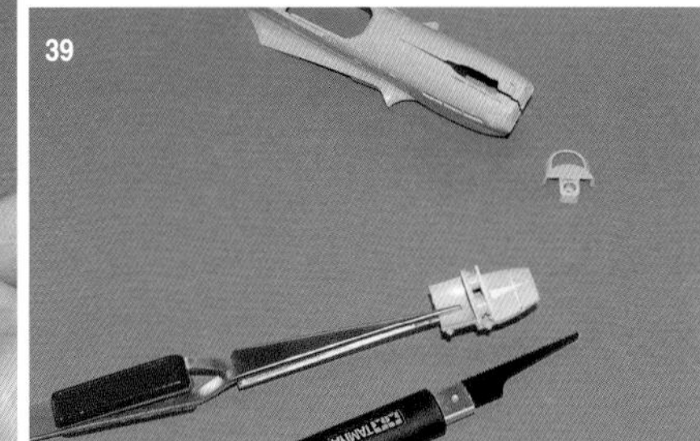

34 to 36. **We make a saw mark to show the emplacement of the radiator grill that we will add.**

37 to 39. **We use the edge of the scissors to cut out a piece of the Hobbykit grill, specially designed for the emplacement of the air intake.**

40. **We cement at the rear of the groove a shaped plastic support.**

41. **The piece that supports the grill is cut out and shaped from plastic sheet.**

42. **The grill is just clipped into the groove. Cementing with cyanoacrylate fills in the spaces of the grill and spoils the detailing.**

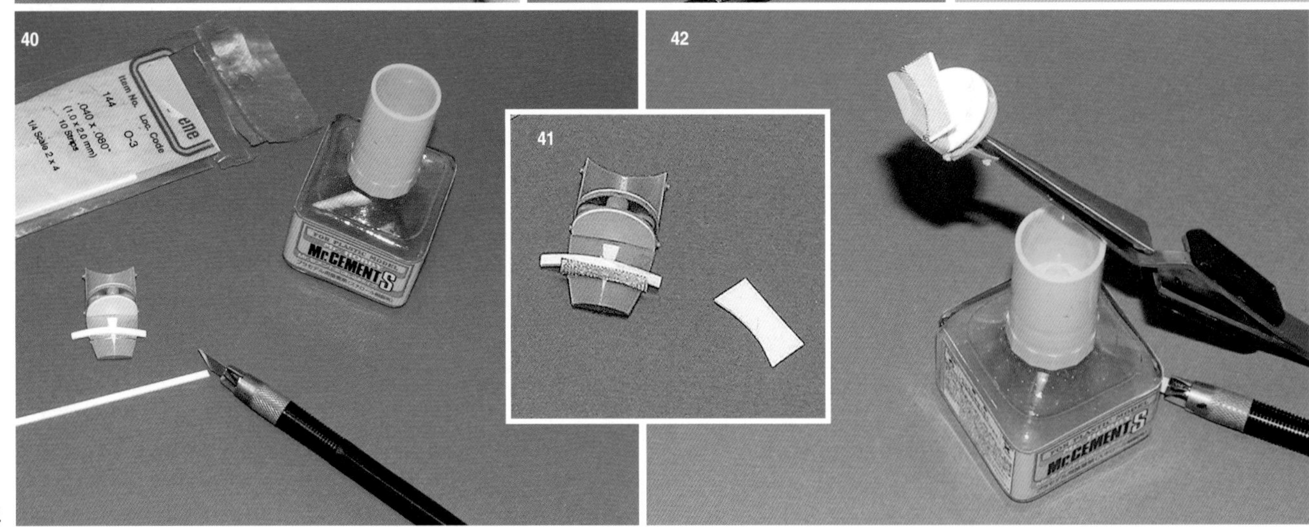

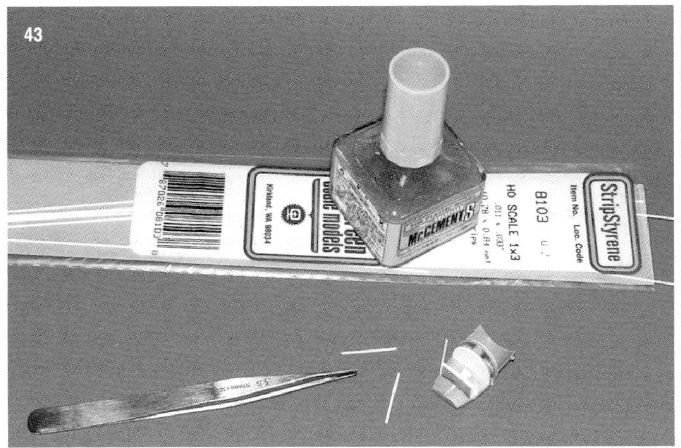

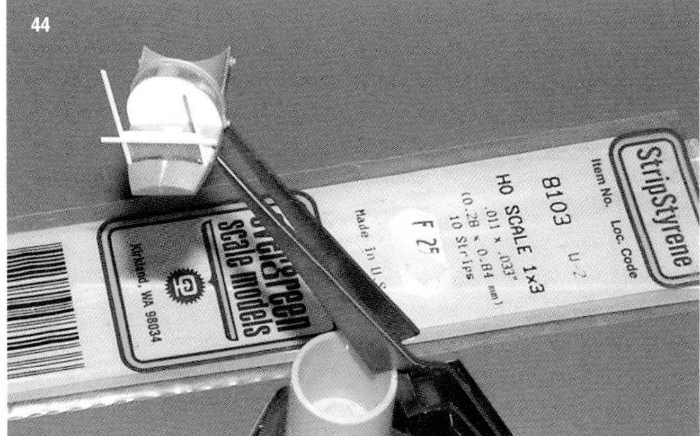

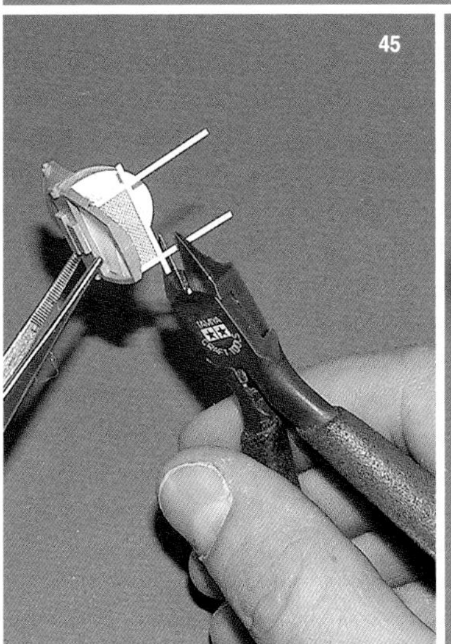

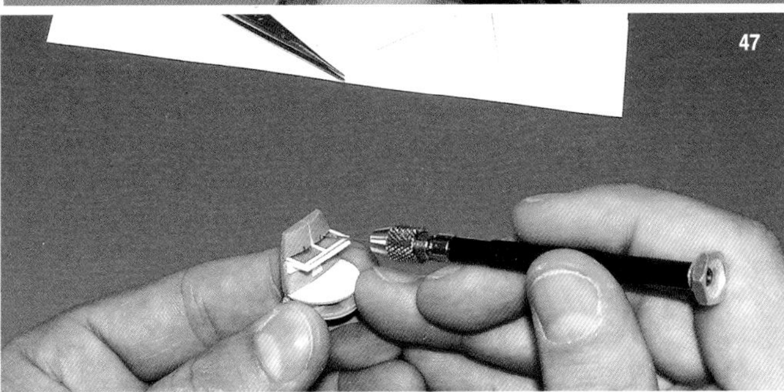

43 and 44. **A styrene strip framework holds the grill in place.**

45 and 46. **Any excess styrene strip is carefully removed.**

47 to 49. **The detail of the air intake is made from scratch.**

50 and 51. **Plastic sheet is slid into the slot that remains after the fuselage halves are cemented. and cemented.**

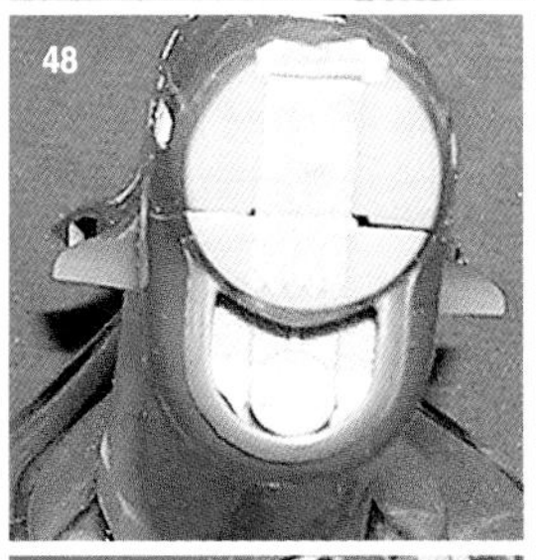

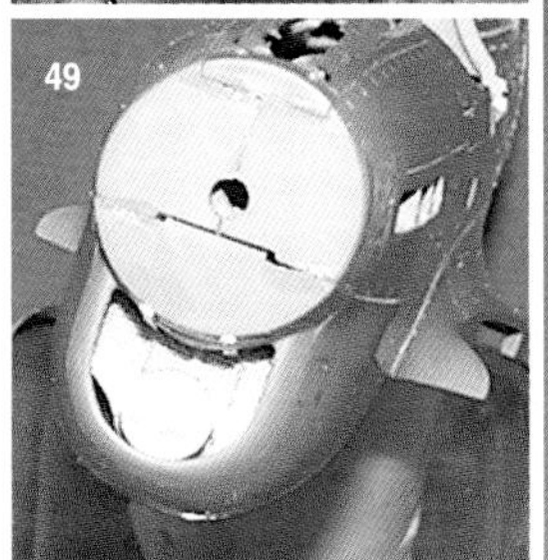

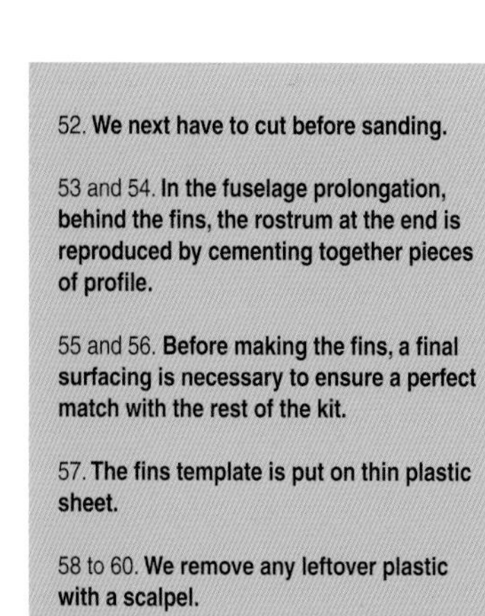

52. **We next have to cut before sanding.**

53 and 54. **In the fuselage prolongation, behind the fins, the rostrum at the end is reproduced by cementing together pieces of profile.**

55 and 56. **Before making the fins, a final surfacing is necessary to ensure a perfect match with the rest of the kit.**

57. **The fins template is put on thin plastic sheet.**

58 to 60. **We remove any leftover plastic with a scalpel.**

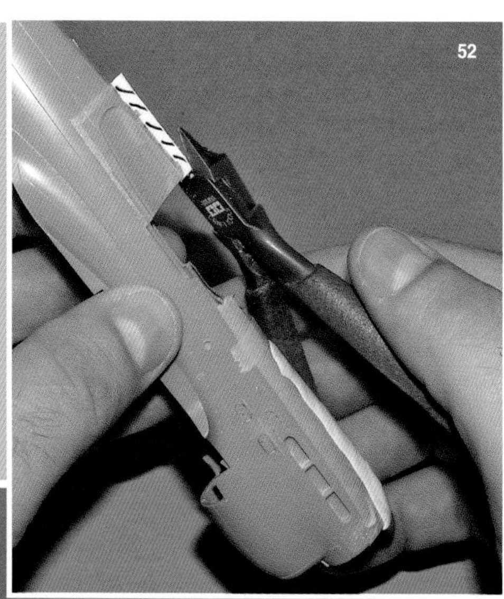

52

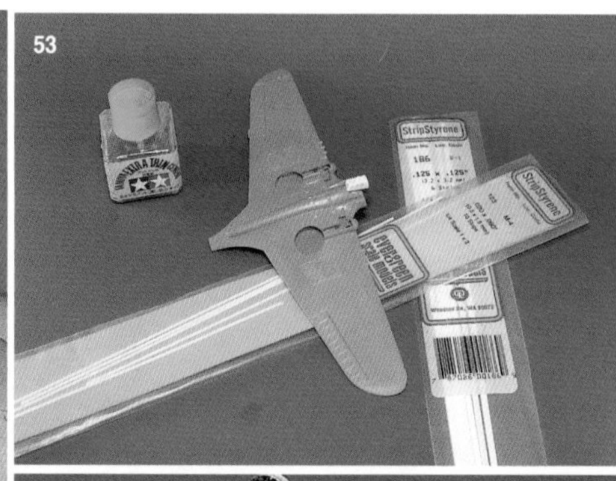

53

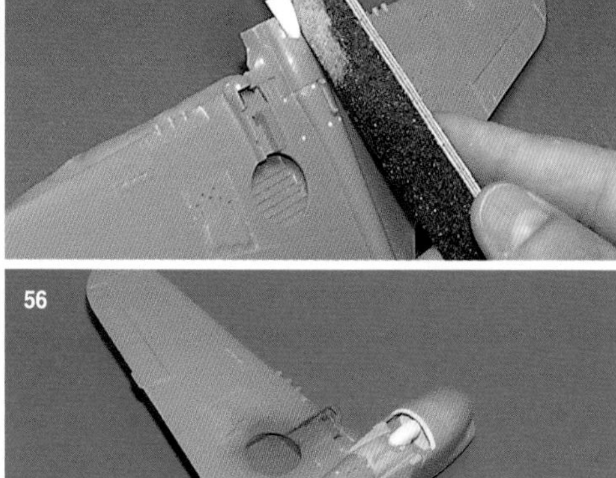

54

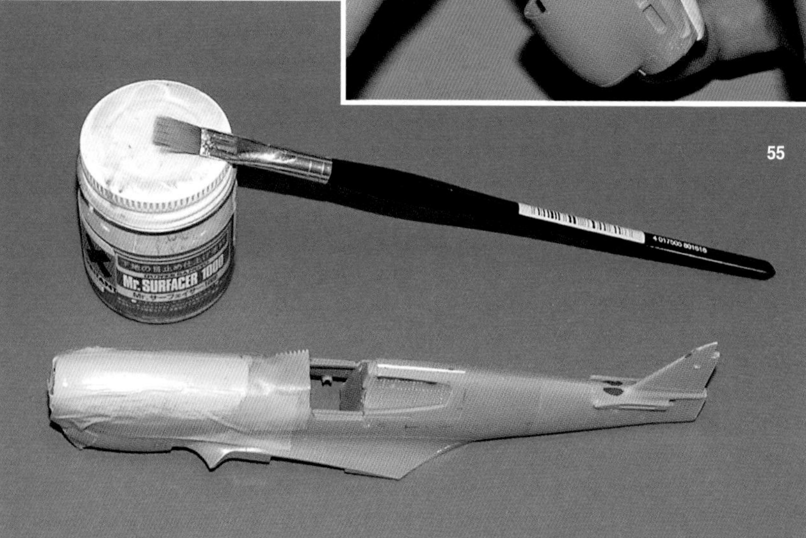

55

56

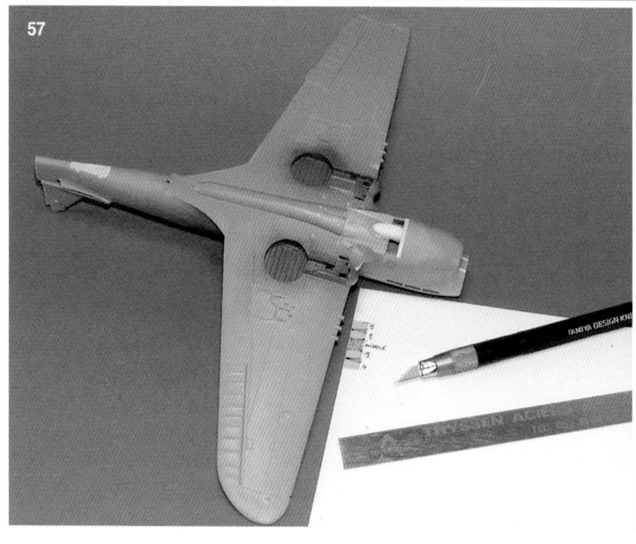

57

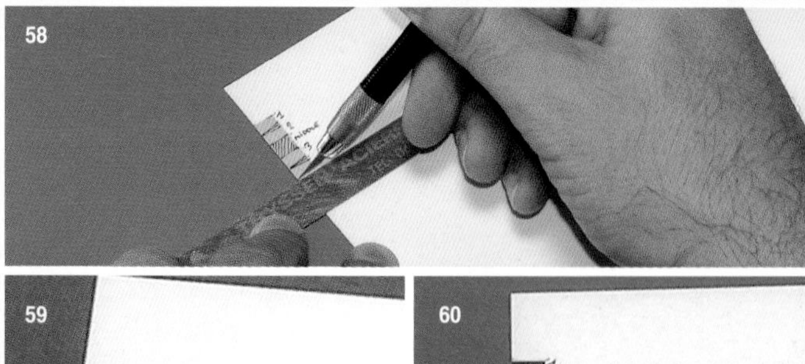

58

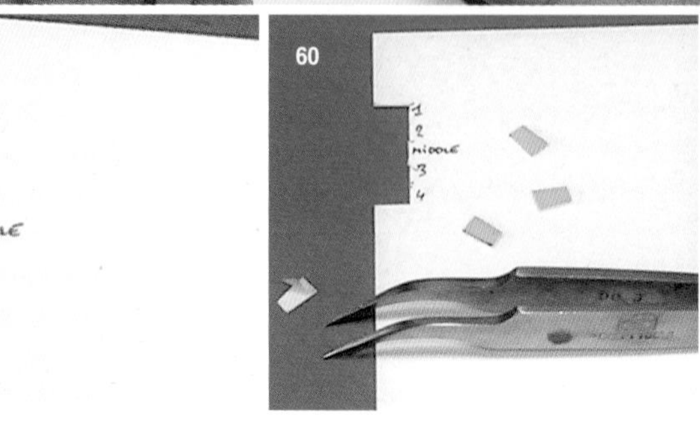

59

60

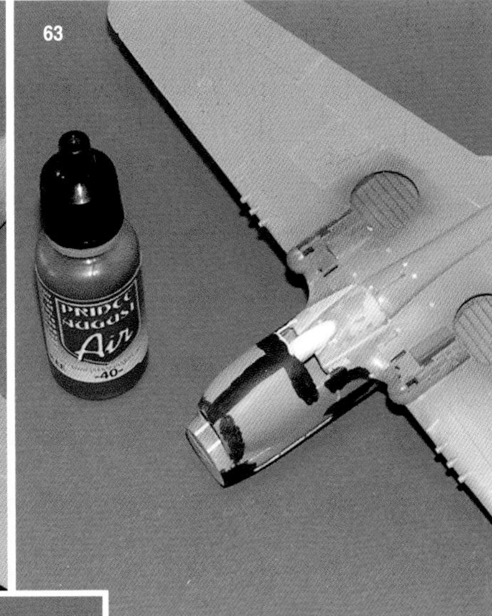

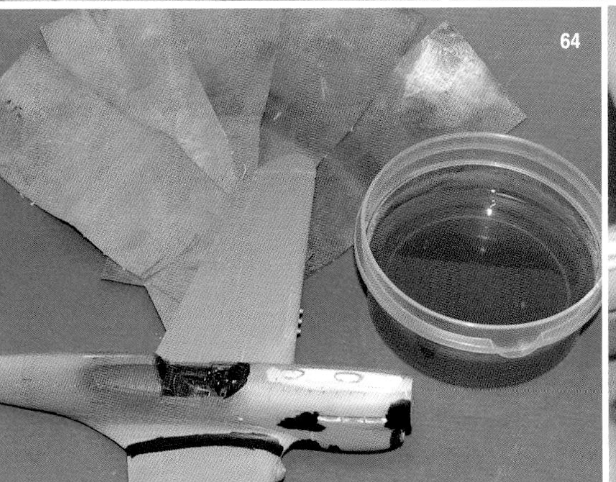

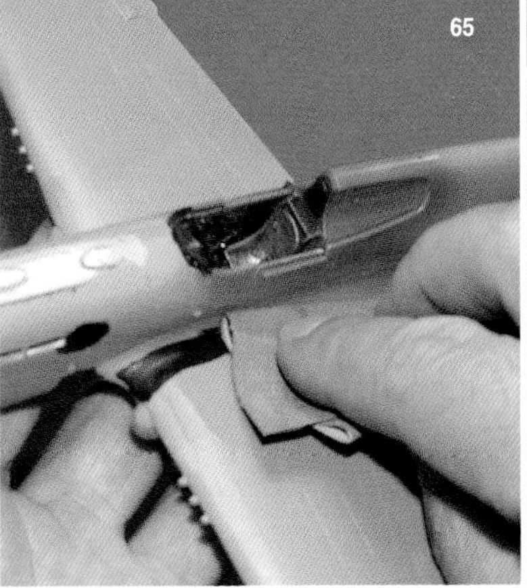

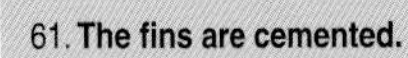

61. **The fins are cemented.**

62 to 65. **A last bit of sanding before starting the re-engraving at the front of the plane.**

66. **The re-engraving materials are ready, the lines are traced.**

67. **The plastified adhesive tape is used as a guide, we go over the lines with a point, counting each passage.**

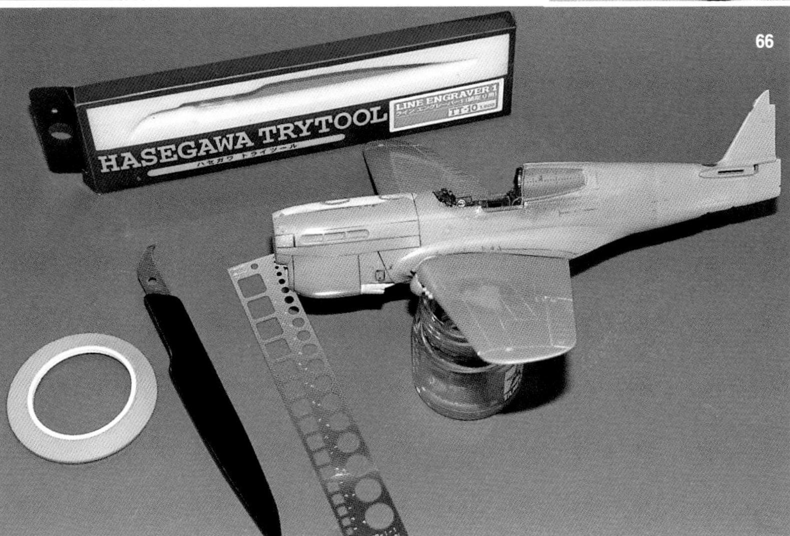

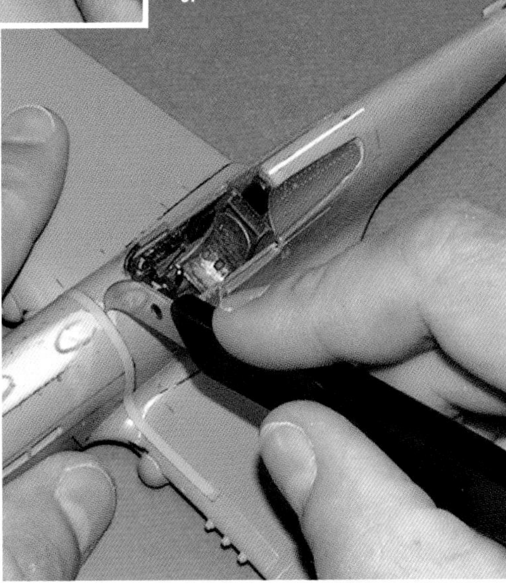

1. Decoration

This aircraft is the personal mount of the famous Commandant Rozanoff, then leader of the GC II/5 'Lafayette'. Aircraft of this group were of British and American origins, the reason why there were variations mainly between RAF Middle Stone and US Desert Sand.

68

69

68. **The model is ready for painting. We use the American desert shades, *Mister Kit, 61492 - Desert Sand -Dark Earth and 61492 - Azure Blue.***

69. **Before adding the camouflage to the upper surfaces, the canopy is cemented. The frame is painted in *U.S. Interior*, then a French.**

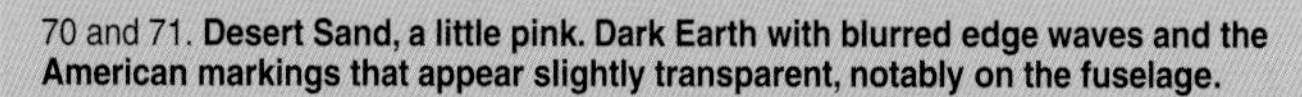

70 and 71. **Desert Sand, a little pink. Dark Earth with blurred edge waves and the American markings that appear slightly transparent, notably on the fuselage.**

70

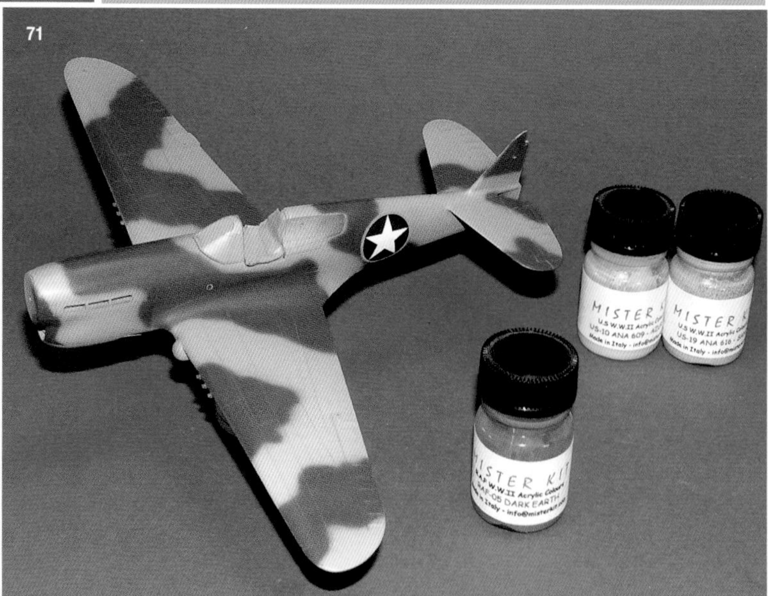

71

13

MADKOT
114
13

114
13
MADKOT

435
13

CURTISS P-40N

PAINTING AND WEATHERING

1/48 **SCALE BUILDING**

by Olivier SOULLEYS

CUTTING **E**DGE, **A**IRES, **CMK**, **E**DUARD, **Q**UICKBOOST, **U**LTRACAST are some of the most important accessory manufacturers on the model market today. Far from the arms race, to the extent that we can no longer shut the box because it is so full of superb improvement kits (!), we suggest making the model 'straight from the box', where the focus is on the painting.
This exercise in style, with great attention being paid to detail in the painting, achieves a quick result and, when all is said and done, a very pleasant one all the way through the work.

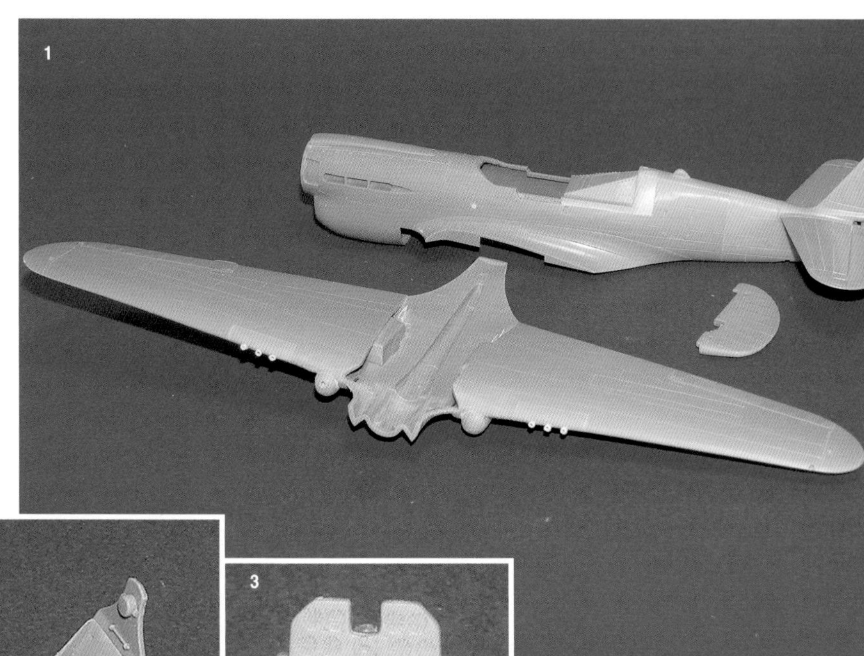

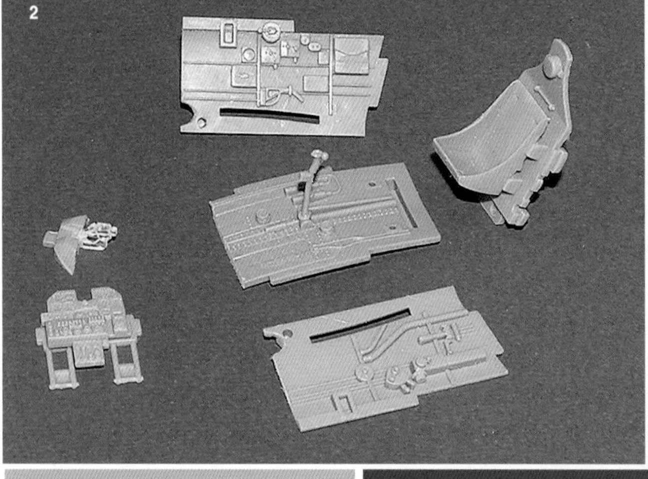

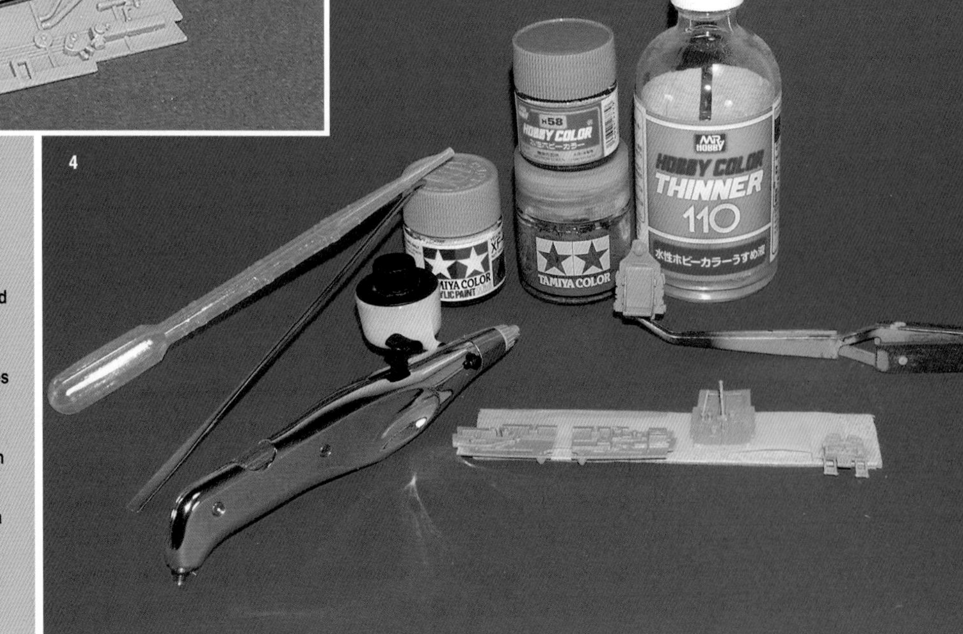

1. We begin by assembling the cell and the wing, having noticed the impressive surface detail for 1/48.

2. The parts of the cockpit are cleaned and imperfections dealt with. Hasegawa has done a good job of reproducing the 'V' shape of the floor which in reality constitutes the top of the wing which goes through the cockpit.

3. The dashboard is fixed to the pedals of the rudder bar, which makes painting with tweezers easier.

4. These elements are fixed on the Tamiya masking tape nice and flat on a piece of cardboard.

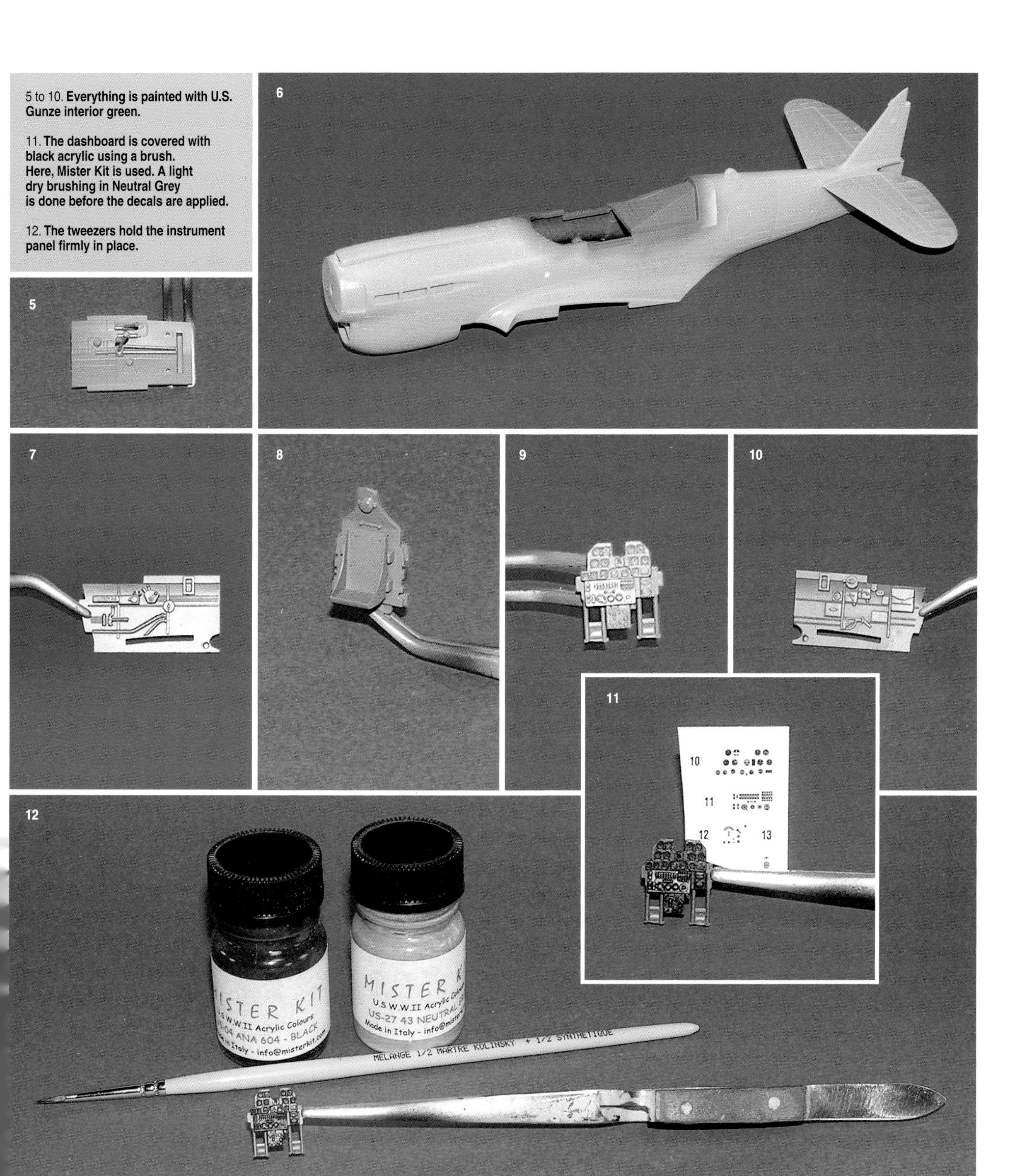

5 to 10. **Everything is painted with U.S. Gunze interior green.**

11. **The dashboard is covered with black acrylic using a brush. Here, Mister Kit is used. A light dry brushing in Neutral Grey is done before the decals are applied.**

12. **The tweezers hold the instrument panel firmly in place.**

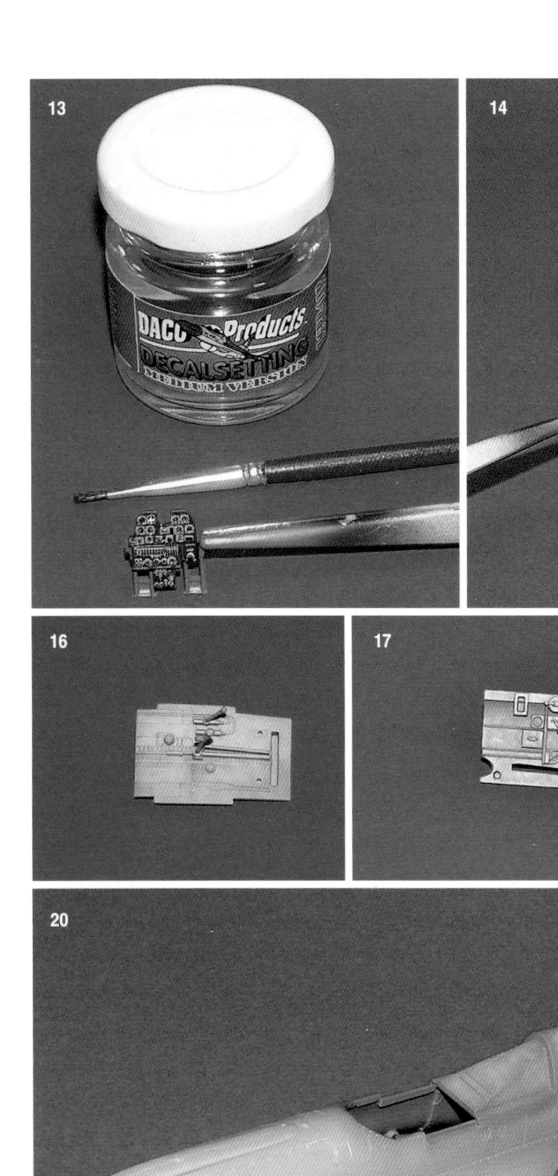

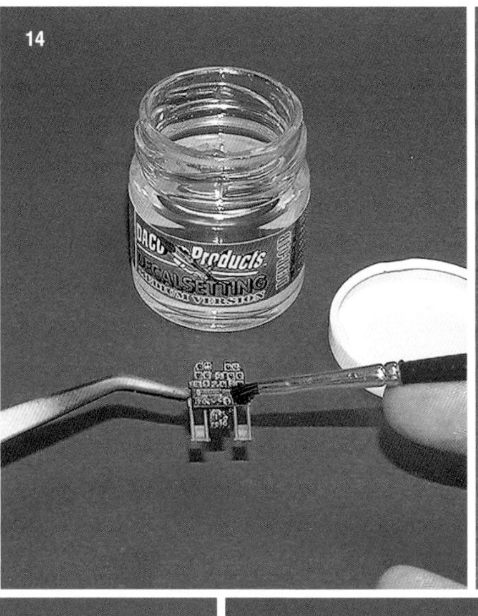

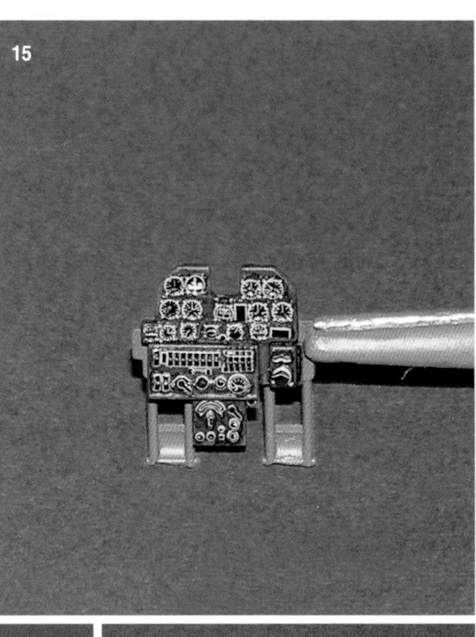

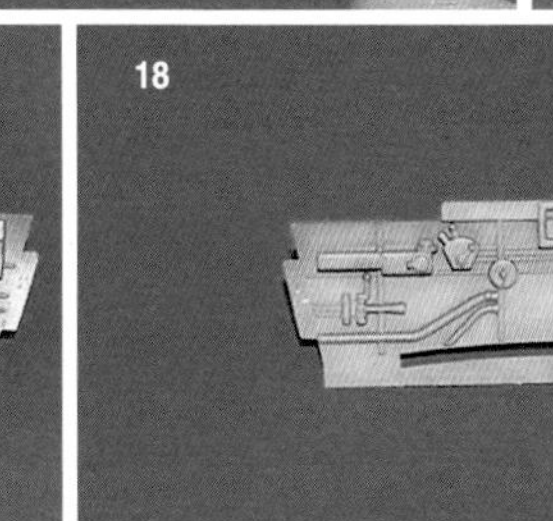

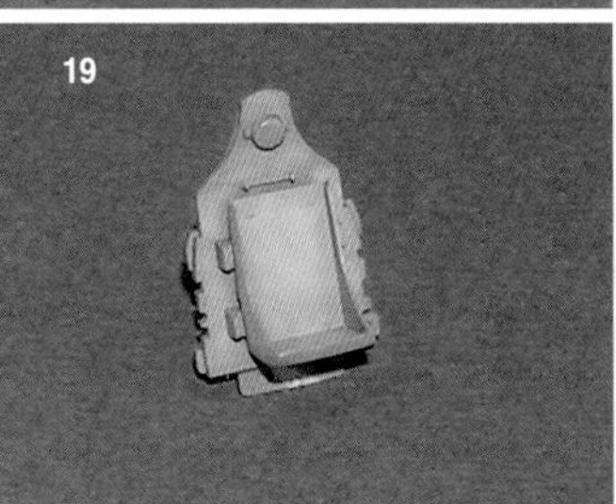

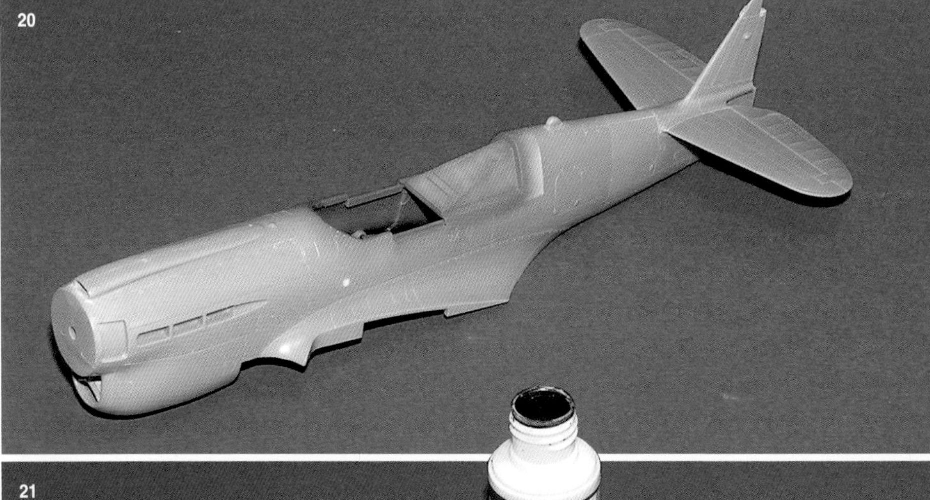

13. **Firstly, all the decals are applied.**

14 and 15. **Lastly, Daco Strong with a stiff paintbrush makes the decals fit the shape as closely as possible.**

16 to 20. **With the equivalent of an extremely diluted, so non-covering, yellow zinchromate (Tamiya XF-4), the green interior can be irregularly shaded.**

21 to 26. **An olive green acrylic is patted on with a foam pad to reproduce an irregular pattern on the surfaces.**

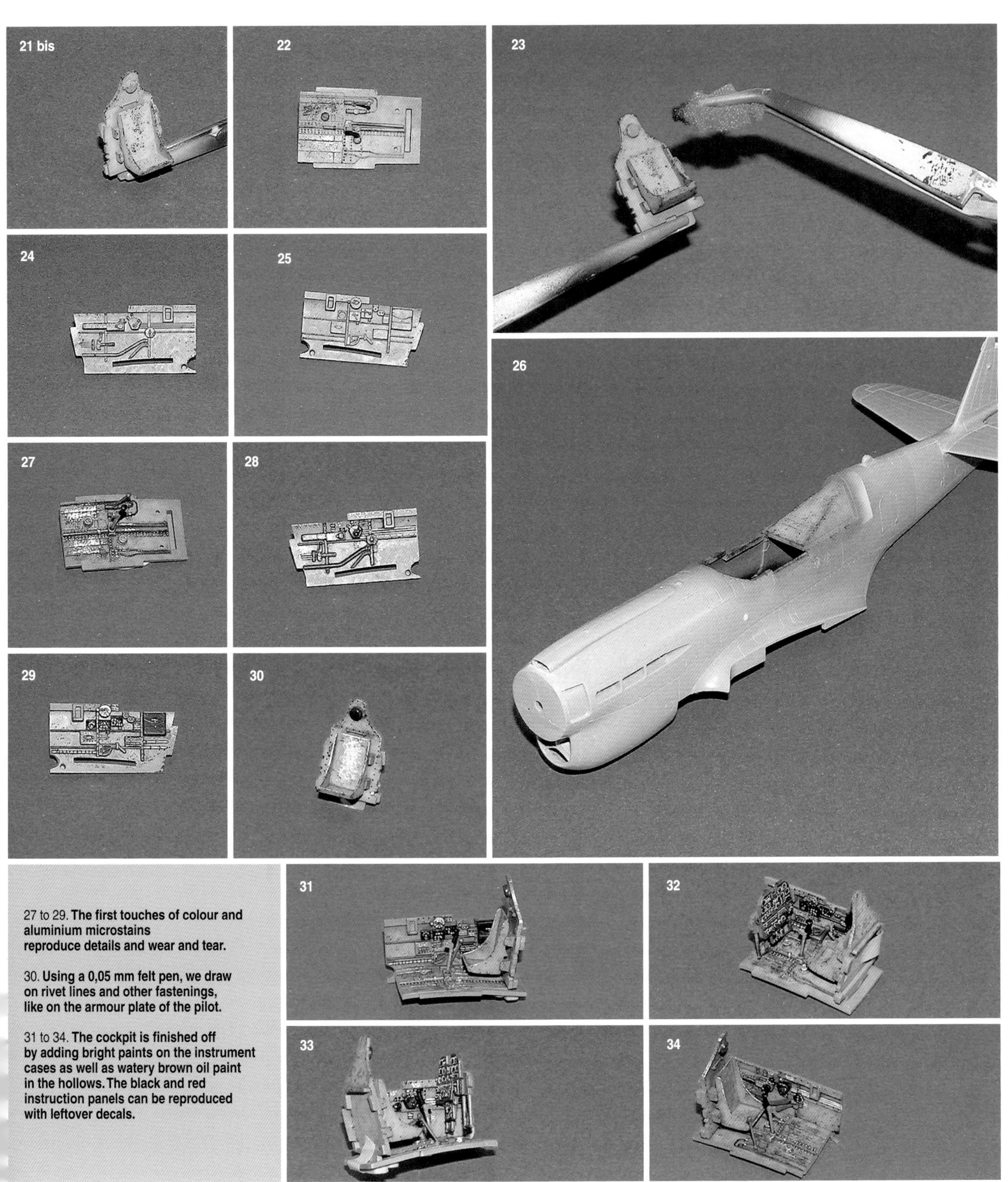

27 to 29. **The first touches of colour and aluminium microstains reproduce details and wear and tear.**

30. **Using a 0,05 mm felt pen, we draw on rivet lines and other fastenings, like on the armour plate of the pilot.**

31 to 34. **The cockpit is finished off by adding bright paints on the instrument cases as well as watery brown oil paint in the hollows. The black and red instruction panels can be reproduced with leftover decals.**

35. The canopy is cemented and masked with Tamiya tape, using a new blade to cut out.

36. In order, a layer of interior green is applied on the canopy, followed by a black one, to make sure that the posts are opaque.

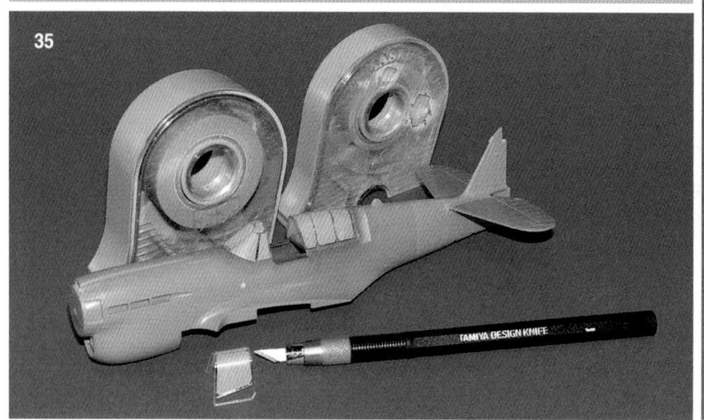

1. Camouflage and markings

An 80th Fighter Squadron plane based in India in the spring of 1944 was chosen. On the menu we therefore have sand dust, earth, a red-hot sun and rain, to make matters worse. Moreover, this machine has been subjected to several complete panel replacements as the front left cowl shows, replaced by one with nose art of a shark mouth. The emblematic skull of the 80th FG remains intact on the right-hand side.

Our weathering method has to respect, above all, the climatic conditions.

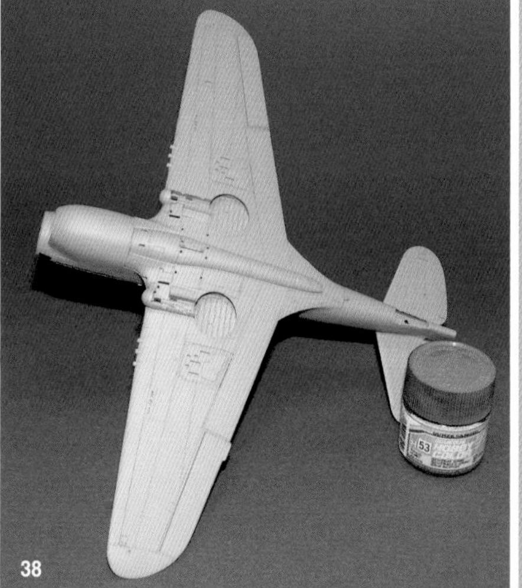

37. Brown-black preshading is applied only on the plane's lower surfaces.

38. Transparently, it appears under the Neutral Grey.

39. With the same Neutral Grey, this time lightened with mat white, the surfaces are 'whipped'more thoroughly, the control surfaces are marked out with Post-its.

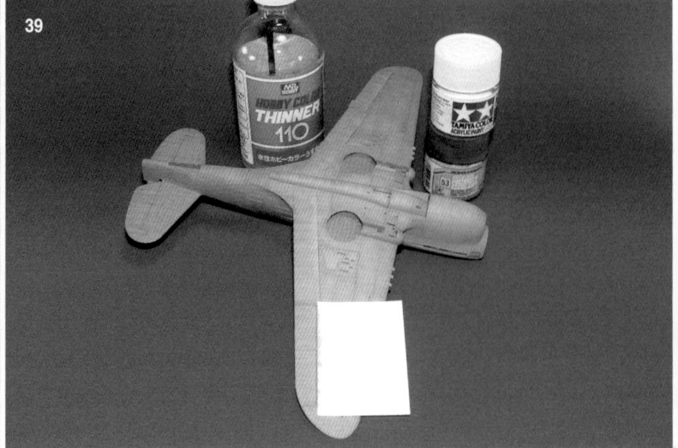

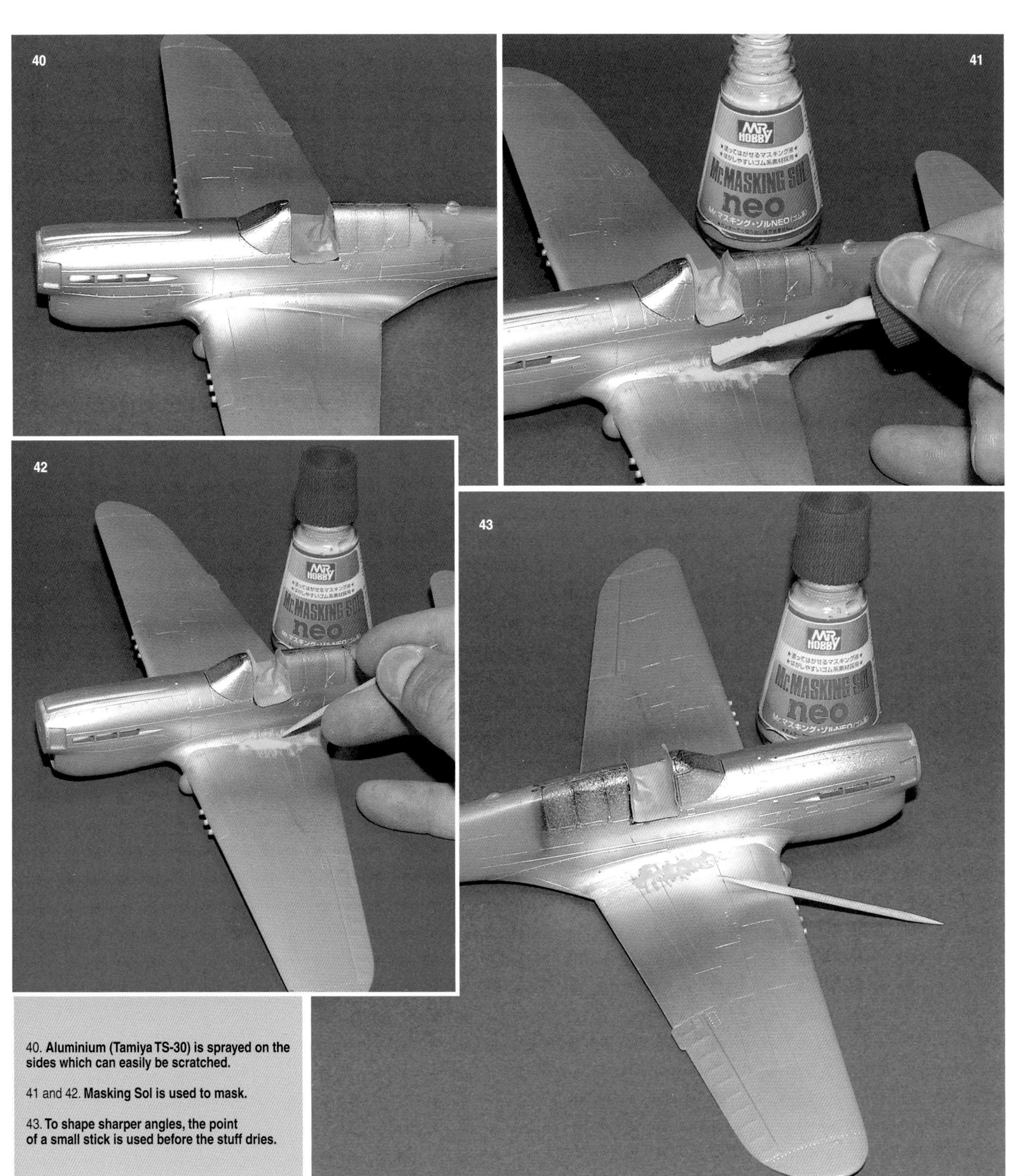

40. **Aluminium (Tamiya TS-30) is sprayed on the sides which can easily be scratched.**

41 and 42. **Masking Sol is used to mask.**

43. **To shape sharper angles, the point of a small stick is used before the stuff dries.**

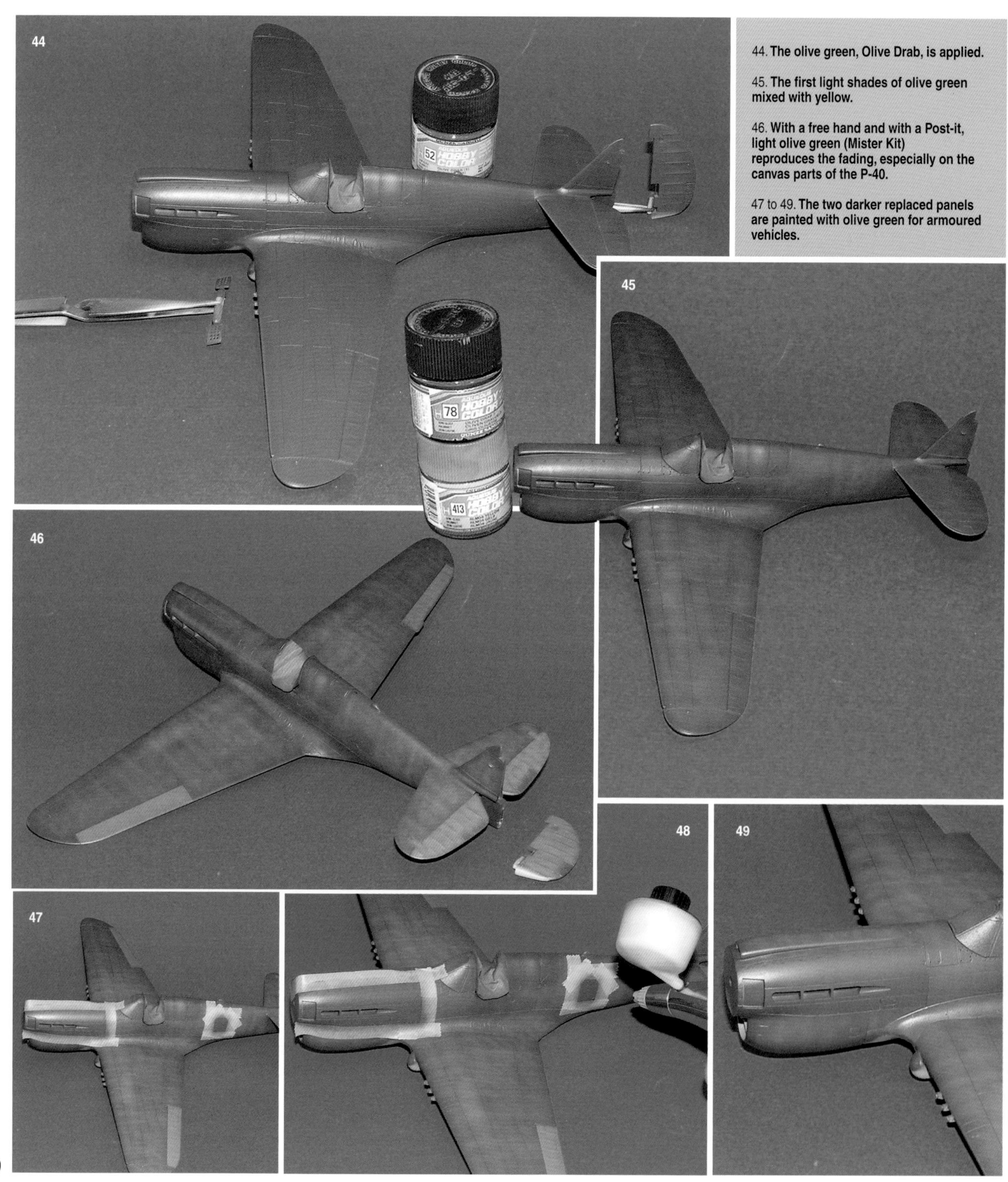

44. The olive green, Olive Drab, is applied.

45. The first light shades of olive green mixed with yellow.

46. With a free hand and with a Post-it, light olive green (Mister Kit) reproduces the fading, especially on the canvas parts of the P-40.

47 to 49. The two darker replaced panels are painted with olive green for armoured vehicles.

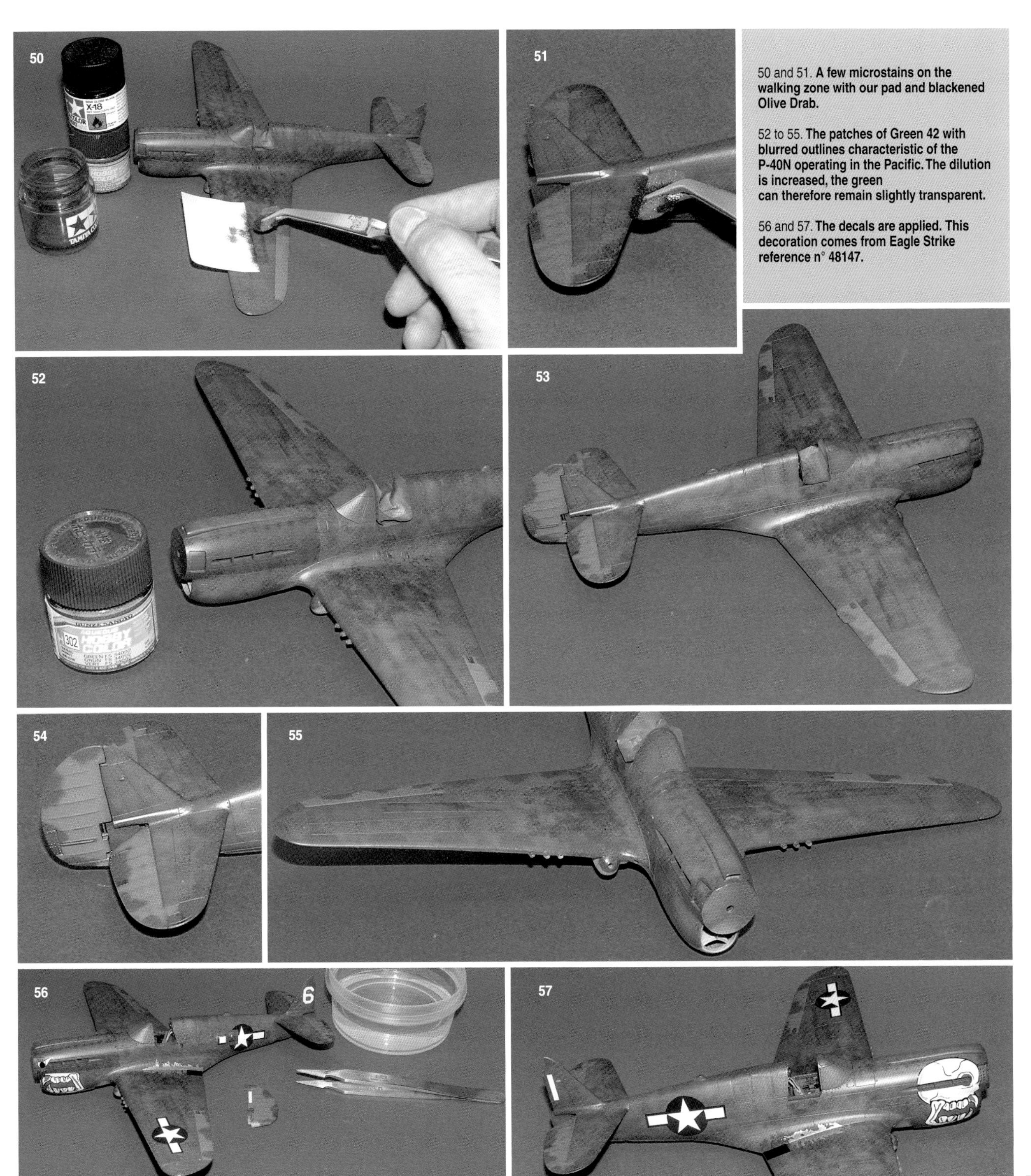

50 and 51. **A few microstains on the walking zone with our pad and blackened Olive Drab.**

52 to 55. **The patches of Green 42 with blurred outlines characteristic of the P-40N operating in the Pacific. The dilution is increased, the green can therefore remain slightly transparent.**

56 and 57. **The decals are applied. This decoration comes from Eagle Strike reference n° 48147.**

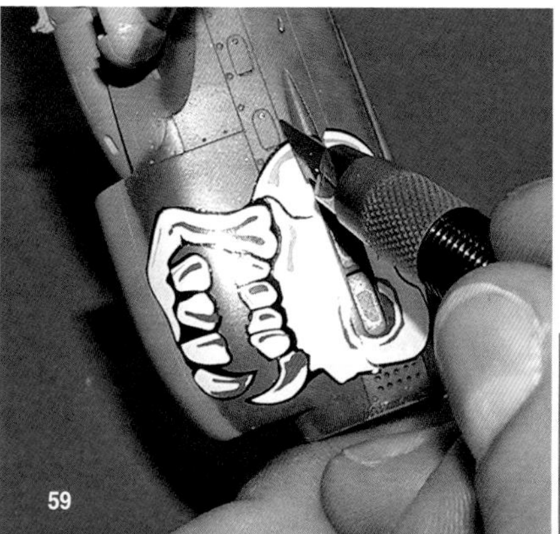

58 and 59. **Once the decals are dry, a scalpel is used to cut the motifs covering the structure lines.**

60 to 63. **The softener enables the decals to seal themselves by fitting snugly into the previously cut structure lines.**

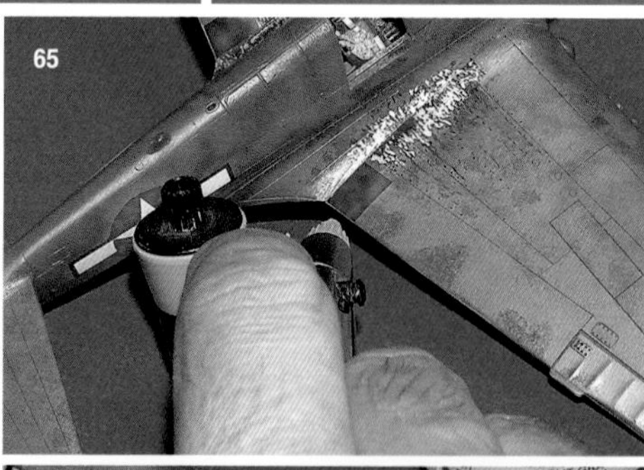

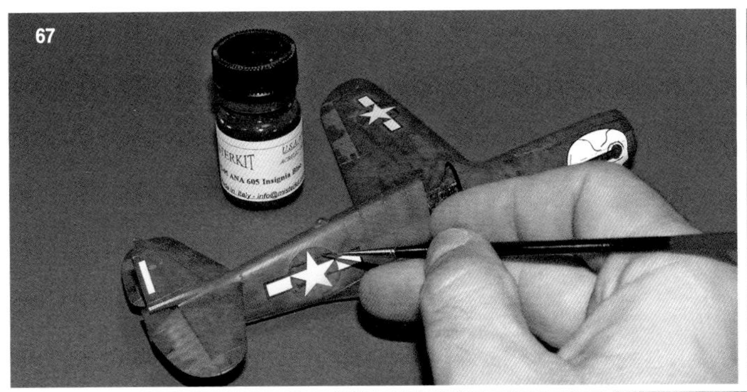

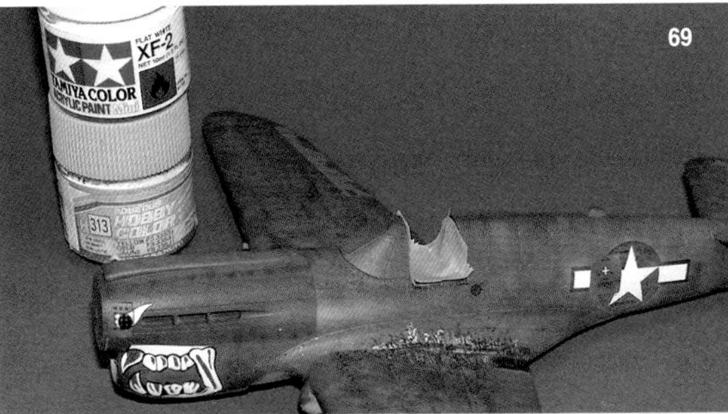

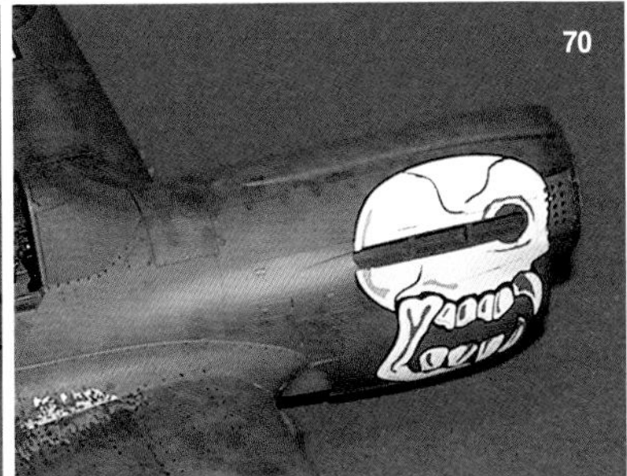

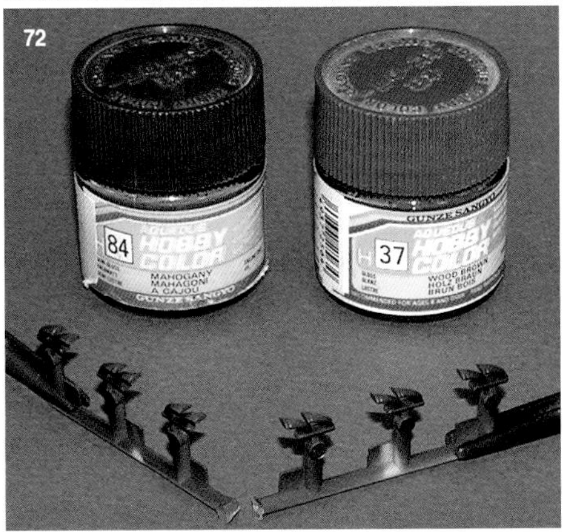

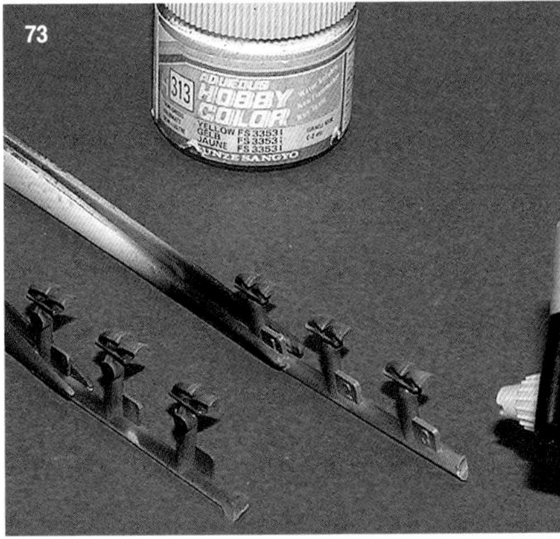

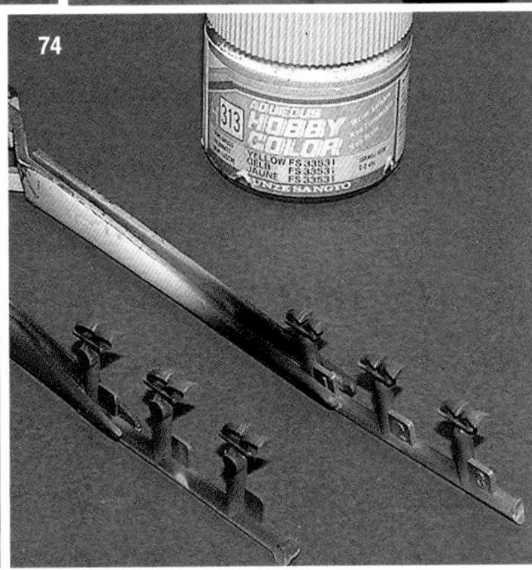

64 to 66. The motifs are lightened with very diluted light grey.

67 and 68. The motifs are re-done with a fine brush and the base shade of the motif.

69 to 71. The beige exhaust fumes, white in the centre. On the left side, the cowl has been changed, so we are careful to reproduce the break in the exhaust fumes, which are less visible on the replaced panel.

72 and 74. The exhaust line is painted in dark brown and lighter on the end of the pipes.

75 to 77. Instead of making a hole, black is painted on the cold set, and using a brush, the edge is lightened to simulate a hollow.

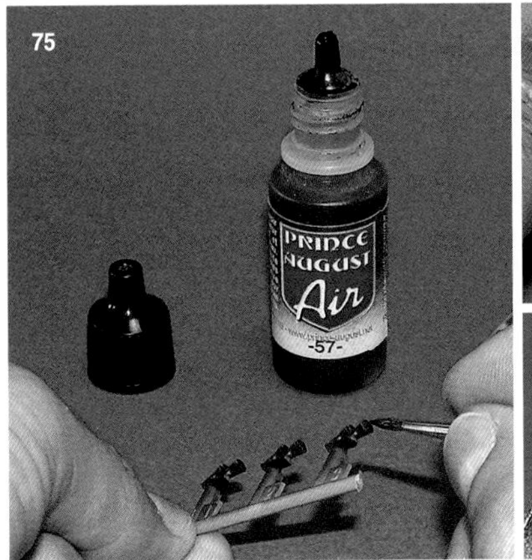

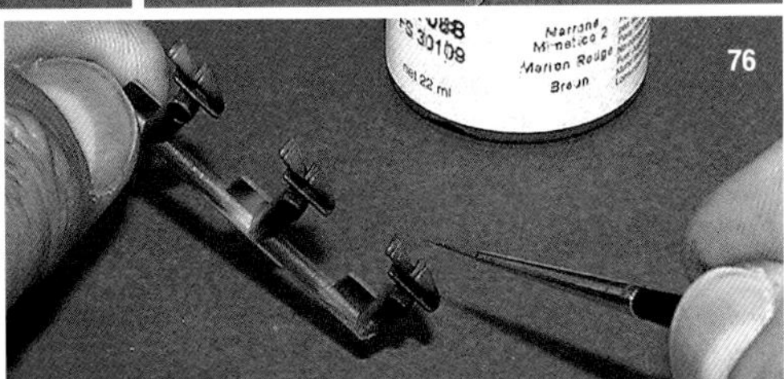

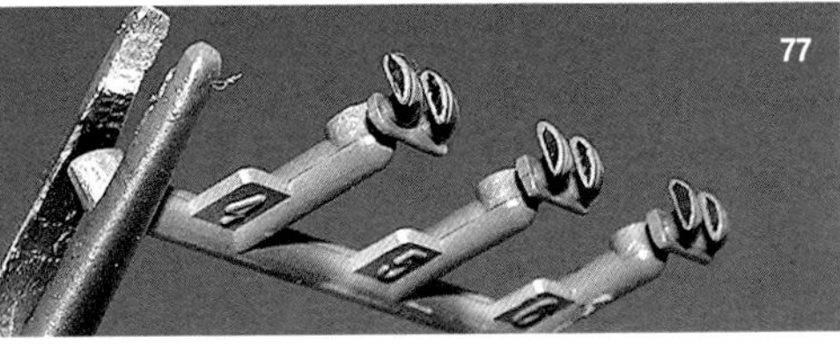

78. The shades d'Olive Drab, the microstains can be seen. Moreover, the aluminium shows through, rubbed away by feet.

79. The spinner and the blades are sprayed with an undercoat of matt white

80. The spinner is painted in bright red.

81. The ends of the blades are painted in yellow without delimitation.

82. The end of the blades is masked.

83. A black satin is sprayed.

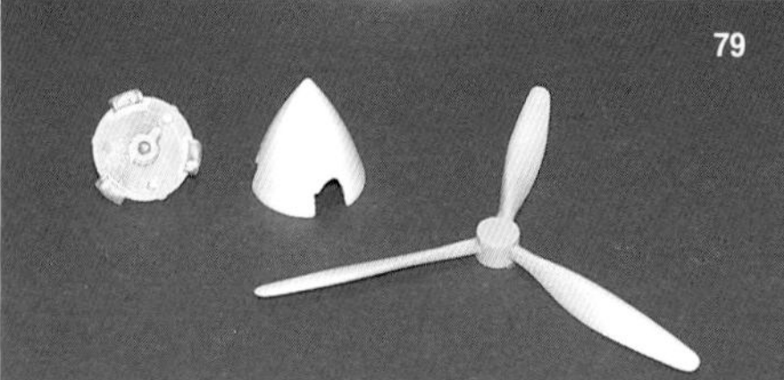

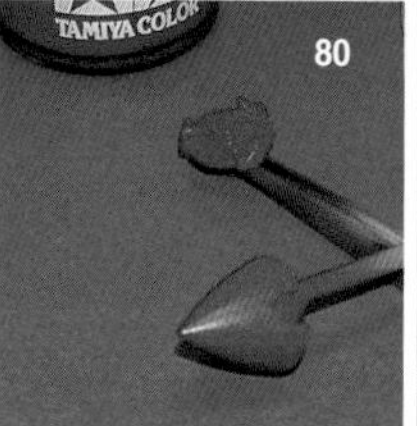

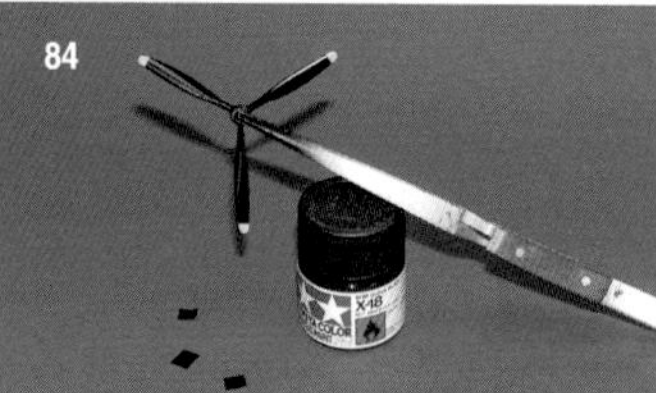

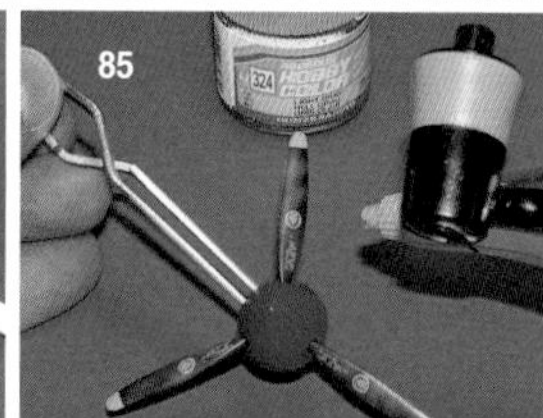

84. The masking tape is removed.

85. A thin layer of grey lightens the blades.

86. The propeller is painted, black oil paint is prepared.

87 and 87bis. The traces of maintenance on the blades.

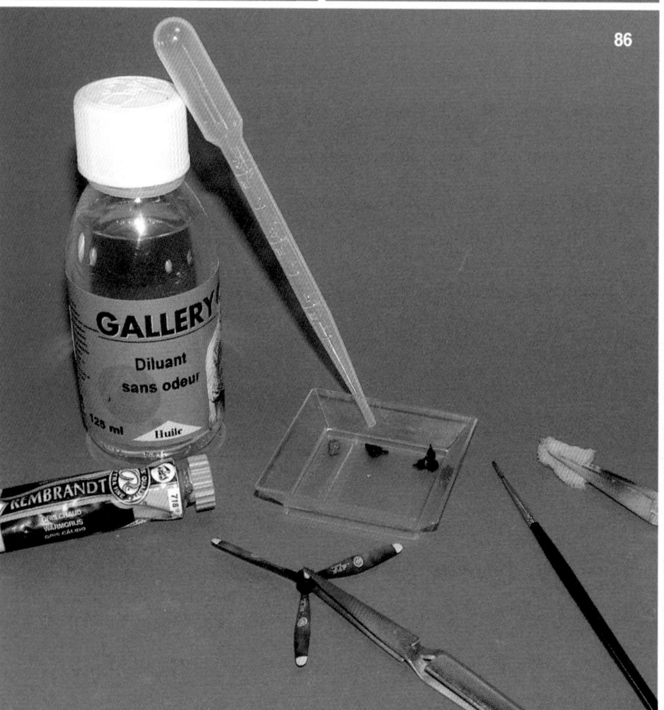

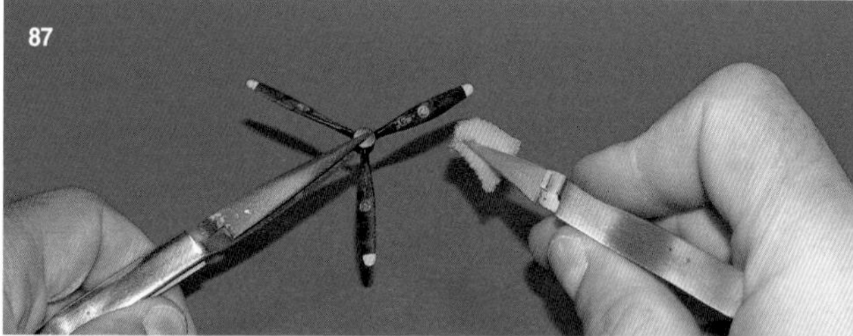

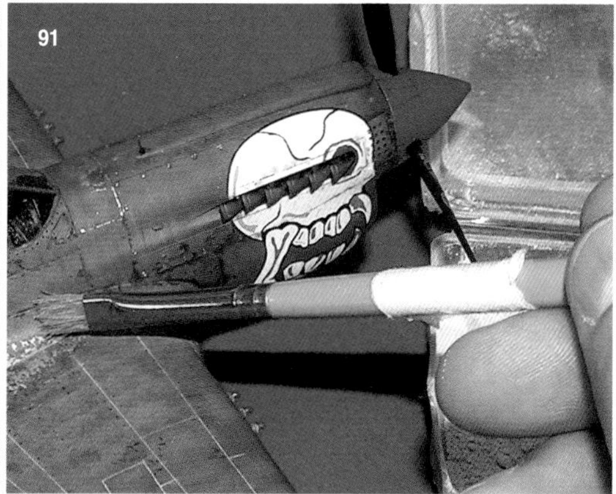

88 and 89. **The clamping pitch of the propeller blades leaks and spits oil. The joint of the propeller cone is marked with an acrylic fine brush, the drips are spread towards the back.**

90 to 92. **The decorative earth appropriate to the theatre of operations is applied with a hard brush. The air direction on the wings must be respected. The wheels get the same treatment especially on the exterior side of the tyres where sand accumulated.**

93 and 94. **The red caps of the machine guns bring a touch of extra colour.**

95. **Soiling is reproduced using blackish oil continuing the line of the cartridge case ejectors and shaded off with a cloth.**

96. **The tail wheel is turned. The fuel cap is painted in red. The fuel stains on the side of the fuselage are bluish and reproduced with a water-colour pencil.**

97. **The dissymmetry of the front cowl is noticeable with the different Olive Drab and the decoration modified by the change in the left motor cowl.**

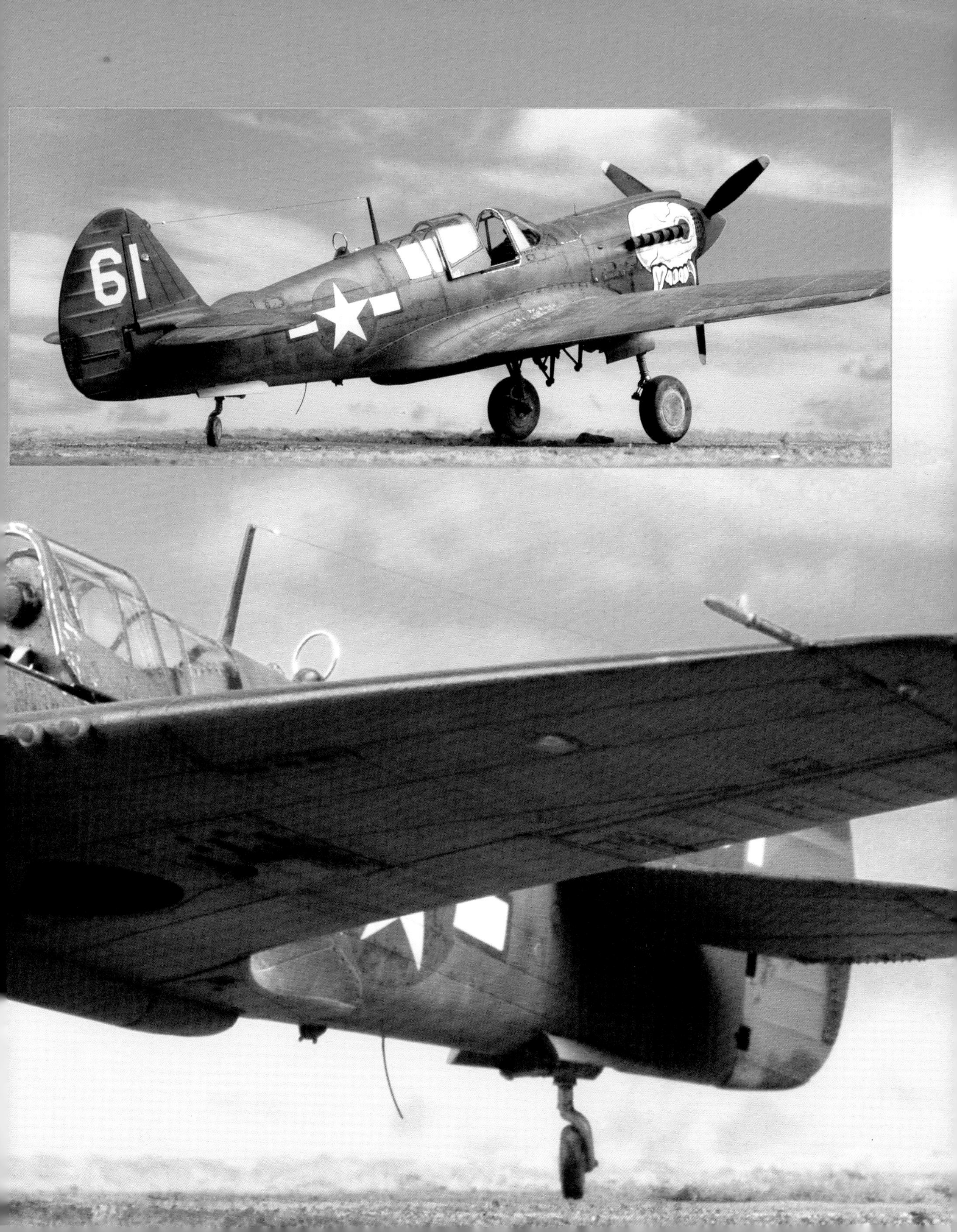
61

61

ACKNOWLEDGEMENTS

The authors would like to thank
Sophie Manrot, Fred CC, André Joubert, Jérôme Monier, Philippe Jouannet, and also Sylvie, Marie and Léa
for their help in the realisation of this book.

Supervision Dominique BREFFORT - Translation Lawrence BROWN.
Design, creation and production Jean-Marie MONGIN and Antoine POGGIOLI,

ISBN: 978-2-35250-065-0

Publisher's number: 35250

Un ouvrage édité par
HISTOIRE & COLLECTIONS
SA au capital de 182938, 82 €
5, avenue de la République
F-75541 Paris Cedex 11
France
Fax 0147005111
www.histoireetcollections.fr

This book has been designed, typed, laid-out and processed by *Histoire & Collections*, fully on integrated computer equipment.

Printed by Zure, Spain, European Union.
August 2008